Growing Resistance

Canadian Farmers and the
Politics of Genetically Modified Wheat

GROWING RESISTANCE

Emily Eaton

UMP
University of Manitoba Press

University of Manitoba Press
Winnipeg, Manitoba
Canada R3T 2M5
uofmpress.ca

Printed in Canada
Text printed on chlorine-free, 100% post-consumer recycled paper

16   15   14   13   1   2   3   4   5

Cover design: David Drummond
Interior design: Jessica Koroscil

Library and Archives Canada Cataloguing in Publication

Eaton, Emily, 1980–
Growing resistance : Canadian farmers and the politics of genetically
modified wheat / Emily Eaton.

Includes bibliographical references and index.
Issued also in electronic formats.
ISBN 978-0-88755-744-6 (pbk.)
ISBN 978-0-88755-435-3 (PDF e-book)
ISBN 978-0-88755-440-7 (epub e-book)

1. Wheat—Genetic engineering—Canada. 2. Transgenic plants—Political
aspects—Canada. 3. Farmers—Political activity—Canada. 4. Wheat—Prairie
Provinces. 5. Wheat trade—Prairie Provinces. I. Title.

SB191.W5E28 2013   633.1'10971   C2012-908121-3

The University of Manitoba Press gratefully acknowledges the financial
support for its publication program provided by the Government of Canada
through the Canada Book Fund, the Canada Council for the Arts, the Manitoba
Department of Culture, Heritage, Tourism, the Manitoba Arts Council,
and the Manitoba Book Publishing Tax Credit.

FSC
www.fsc.org
MIX
Paper from
responsible sources
FSC® C016245

Contents

Tables &
Illustrations

TABLES

Table 1
Organizations involved in the 31 July 2001 coalition to stop the
introduction of Roundup Ready (RR) wheat   / 14

Table 2
Major importers of Canadian wheat   / 15

Table 3
Exports by country of destination for wheat, durum wheat,
and wheat flour, 2001–2002 crop year   / 15

Table 4
Seeded area (acres) in the prairie provinces   / 57

Table 5
Percentage of farms growing wheat in the prairie provinces   / 57

Table 6
Average farm price in the prairie provinces   / 78

Table 7
Annual returns for wheat and canola   / 78

ILLUSTRATIONS

Acknowledgements

Thanks go first and foremost to the participants in this research who gave up their precious time to this project and answered difficult questions from a probing outsider. My mom (Valerie Veillard) and my dad (Bob Eaton) provided me with a home and access to a car during my field work and are significantly responsible for the success of my field work. They instilled in me a keen sense of social justice from day one and for that I am most grateful. Mom also helped me rework parts of the dissertation into a more accessible book.

This book is based on my doctoral dissertation, which I wrote from within a wonderful community of friends and academics at the University of Toronto. Scott Prudham played a huge role as my supervisor. He gave me rigorous yet generous and encouraging feedback on written work and contributed much to my broader intellectual development. I am lucky to still call him a friend. I am also grateful to my supervisory committee—Deborah Leslie, Michael Bunce, Emily Gilbert, and Josée Johnston—and my external examiner—Gavin Bridge—for the important feedback and guidance they gave me.

During my time in Toronto I benefited immensely from amazing friends and colleagues. Amy Siciliano, Vanessa Mathews, Jenn Ridgley,

Patrick Vitale, Roger Picton, Jaume Franquesa, Marion Traub-Werner, Lisa Freeman, Paul Jackson, Kate Parizeau, Suzanne Mills and Zsolt Szekely. I hope we can continue reading, socializing, analyzing, debating, and organizing no matter where our lives take us.

Here in Regina this book was made possible by the security and research time provided by a tenure-track job (shamefully fewer and further between these days). The editors, publishers, and marketers at the University of Manitoba Press were a pleasure to work with—so thorough and enthusiastic. Thanks for convincing me this was a story worth telling. Two anonymous reviewers provided kind suggestions, and Celia Braves skilfully crafted the index.

My 'love bug' Simon Enoch and our Booker dog and Twenty cat continue to provide such a loving and fulfilling life. Simon has been so generous, not just with his love, but also by reading and commenting on chapters and sustaining an ongoing conversation about the difficult world we live in. These conversations about what is wrong and what is to be done are what keep me going.

Preface

In 2001, a unique coalition of organizations stood together against the introduction of genetically modified wheat to Canadian farms. This study of that movement is supported by a series of interviews I conducted with representatives of these organizations, employees of Canadian government agencies (including plant breeders, scientists, and regulators), representatives from biotech lobby groups, industry organizations, Monsanto Canada, and members of farm organizations that publically supported the introduction of Roundup Ready (RR) wheat.

I became aware of just how political the controversy over GM wheat had been when I began arranging interviews with agencies and employees of the Canadian government. Some scientists at Agriculture and Agri-Food Canada (AAFC) replied to my messages indicating that their positions as federal scientists meant that it would be inappropriate for them to comment on the politics of GMOs. Federal scientists, I was told, did not enjoy the same academic freedom as their counterparts at universities. Others at AAFC generously accepted the invitation to participate in my research, but they all shied away from my questions about how research agendas are determined and especially about how science might be understood as a contested domain.

Getting an interview with anyone at the Canadian Food Inspection Agency (CFIA), the agency charged with the environmental regulation of plants with novel traits, proved to be the most challenging and frustrating part of the research. Over and over I was told that the CFIA regulates based on sound science; therefore, no one at that organization would have anything to say about the politics of GMOs. More than once, an interview was cancelled because it had not received clearance by the interviewee's supervisor. My consent form was sent to the CFIA's legal department, and when I finally managed to get anyone to speak with me they would not sign their consent or allow the interview to be tape-recorded. It should be noted that these difficulties were encountered in trying to recruit interviewees at all levels of the hierarchy, from field inspectors to those in charge of the Plant Biosafety Office.

In this book I rely quite heavily on the interviews I conducted, and present much of the story of RR wheat through my participants' voices. In addition to these forty-three interviews, I examined all articles pertaining to genetic modification in Western Canada's most prominent weekly farm newspaper, *The Western Producer*, from 2000 to 2006, and attended five public farm meetings.

KEY CONCEPTS

A basic understanding of a few key concepts is essential for reading this book. To begin, the reader should have a sense of what I mean by *neoliberalism* (or *advanced liberalism*). According to geographer David Harvey, "neoliberalism is in the first instance a theory of political economic practices that proposes that human well-being can best be advanced by liberating individual entrepreneurial freedoms and skills within an institutional framework characterized by strong private property rights, free markets, and free trade. The role of the state is to create and preserve an institutional framework appropriate to such practices."[1] The political-economic *theory* of neoliberalism began to be implemented as political-economic *practice* in the 1970s, beginning with the governments of Margaret Thatcher in Britain and Ronald Reagan in the U.S. Neoliberal policy quickly spread across the globe through the 1980s and 1990s. Exactly how it has been implemented has varied across the globe, but as an ideal type neoliberal practice has involved the deregulation of labour and product markets in order to reduce "impediments" to business; privatization of state-owned enterprises and state-provided services; liberalization of trade in goods and capital investment; and a monetarist economic policy that, for example, treats spending on welfare as a cost of production rather than a source of domestic demand. Neoliberal policy not only erodes or "rolls back" the post-war Keynesian welfare state, but it also entails a "roll out" of new state policies. For example, the neoliberal state actively re-regulates the economy and society by fostering public-private partnerships, involving new "stakeholders" in decision-making processes, harmonizing and internationalizing state policy, and "cracking down" on crime, immigration, and welfare "abuse" through new practices of policing and surveillance. As many critics have shown, neoliberal states, while claiming to limit the role of governments, are deeply interventionist in practice.[2]

The neoliberal project is not only about rearranging political economic practice; it is also a process of *subjectification* (it aims to transform the way people understand themselves and the world and, thus, the ways people act in the world). In order to understand the concepts of subjectification and subjectivity it is useful to return to one of the central debates among social scientists concerning the relative importance of what are frequently called "structure" and "agency." Structure and agency are two extremes of a spectrum that is concerned with the potential and the exercise of human action. On one extreme are those who suggest that individual action is dictated by the social structures that govern all individuals (that individuals are completely constrained by the power relations within which they live). This extreme position, which posits that individuals cannot act outside of societal structures, is usually associated with a structural version of Marxism and many post-structuralist bodies of thought. For Marxists, for example, it is impossible to act outside of the structural constraints that capitalist economic relations impose on us all. That contemporary life is structured around markets means that individuals are compelled to sell their labour, to buy their subsistence, and to respect private property. It also means that corporations are compelled to think only about returning profits to their shareholders; this focus on profit systematically shapes and determines all of the operations of the corporation. For post-structuralists, discourses such as those associated with femininity, race, and foreignness shape how individuals come to understand themselves and the world around them. And while discourses may change over time, it is impossible not to be subjected to them. In this perspective, individuals are fully constituted by a series of narratives that tell them who they are and how to be. Just as a female who attempts to resist the discourse of femininity is always understood in relation to that discourse, a farmer who resists the neoliberalization of agriculture is understood in relation to the discourse of prairie farmers that is characterized by individualism, entrepreneurship, and masculinity.

On the opposite pole of the spectrum are those who emphasize the agency of individuals (the capacity of individuals to determine the course of their own lives through a series of conscious decisions). For liberal humanists, for example, social structures are the product of human agency and can therefore be reformed, abolished, or changed through conscious and directed effort. Humans are, thus, authors of their social worlds; they are free agents who act rationally and with intention.

While many post-structuralists understand discourses as constitutive of social life, they also show how discourses change over time and how people resist them in their everyday lives by, for example, "gender-bending," refusing to indicate their "race" on census forms, or pluralizing proper nouns such as America to indicate the diversity of experiences and peoples in the region. For post-structuralists, such everyday resistances are expressions of agency, of people strategically fashioning the course of their lives. The extreme poles of the structure/agency debate are caricatures and are infrequently reproduced in academic writing. Nevertheless, the debate over the relative importance of and emphasis on one pole or the other continues to rage.

Related to debates about the capacity of individuals to fashion their own lives is the concept of subjectivity. A subject, unlike an object, is an agent who acts according to his or her perspectives, beliefs, and desires. Thus, a subject has some degree of agency that is exercised according to human conditions. A subject is also the product of structural relations of power: s/he is subjected to dominant relations of power and ways of being. For example, under neoliberalism, people's ways of being (of understanding themselves and of acting in the world) are shaped by the economic, social, and cultural relations of neoliberalism as a system of governance. In other words, neoliberal relations promote and enforce on individuals a certain way of being (a subjectivity) that often has to do with understanding oneself as personally responsible for one's fate. This includes making the right life choices according to market signals. Neoliberalism shapes our perspectives, beliefs, and desires and thus guides how we act in the world. Neoliberal discourse makes certain policies and explanations seem natural (that farmers should have the "right" to market their products how they see fit) and others seem unfair and constraining (that taxpayers should pay for people too "lazy" to get a job). In this way agency (for example, our capacity to set up collective marketing organizations and progressive taxation structures) is constrained by the parameters of social and economic structures and discourses.

Finally, *positionality* refers to the idea that "where we are located in the social structure as a whole and which institutions we are in and not in have effects on how we understand the world."[3] Most importantly for the arguments in this book, positionality draws attention to the fact that neoliberal subjectivity is reinforced and resisted in uneven ways, by different people, according to their relationships to particular social structures and institutions. In other words, individuals and groups are differently located in relation to various social structures and institutions, and their relative location affects their experiences and their capacity to act in the world. For example, a white, working-class man may experience exploitation in his work relations while at the same time benefiting from white skin and male privilege in his neighbourhood.

In this book, prairie farmers are deliberately presented as subjects, as active agents in the political economy of prairie agriculture, rather than as passive victims of governments and corporations. At the same time, the book also highlights the neoliberal discourses and policies within which farmers have had to wage their struggle. Working within neoliberal governance while seeking to transcend aspects of it has necessitated some fancy footwork, which I hope this book begins to explain.

Abbreviations

AAFC	Agriculture and Agri-Food Canada
APAS	Agricultural Producers Association of Saskatchewan
CBAC	Canadian Biotechnology Advisory Committee
CCC	Canola Council of Canada
CEPA	Canadian Environmental Protection Act
CFIA	Canada Food Inspection Agency
CoC	Council of Canadians
CWB	Canadian Wheat Board
GGGC	Grain Growers' Grain Company
GE	genetically engineered
GM	genetically modified
GMO	genetically modified organism

KAP   Keystone Agricultural Producers

NBAC   National Biotechnology Advisory Committee

NFU   National Farmers Union

PNT   plant with novel traits

RAC   Rapeseed Association of Canada

RR   Roundup Ready

RSC   Royal Society of Canada

SARM   Saskatchewan Association of Rural Municipalities

SOD   Saskatchewan Organic Directorate

WRAP   Wild Rose Agricultural Producers

Growing Resistance

INTRODUCTION

On 10 May 2004, the Monsanto agricultural corporation announced it would "discontinue breeding and field level research of [transgenic] Roundup Ready wheat."[1] It was a drastic decision; Monsanto had already invested many years and resources in breeding, conducting field trials, and advancing this product through the Canadian and American regulatory systems. Citing an extreme reduction in "business opportunities" for this product, Monsanto conceded to the widespread politicization of Roundup Ready wheat in the U.S. and especially in Canada. Interestingly, active and adamant opposition to RR wheat contrasts with the small amount of public opposition to the transgenic herbicide-tolerant canola varieties (including RR) that were introduced in Canada in 1995. In fact, prairie farmers had quite readily adopted those varieties, and by 2005 transgenic herbicide-tolerant canola accounted for 78 percent of all canola grown nationally.[2] The biotechnology industry and Canadian regulators were, thus, somewhat taken aback when in July 2001 a broad range of farm, consumer, environmental, health, and industry organizations joined forces to publicly voice their opposition to RR wheat.

Why is there such a discrepancy between the widespread adoption of RR canola on the prairies in the mid-1990s and the strong opposition to RR wheat by the same farmers less than a decade later? This is the question that has animated my research and this book. Initially I turned to the cost-benefit analyses of agricultural economists for insight.[3] Such analyses considered a wide range of factors, including the potential damage of herbicide-tolerant weeds, losses due to the contamination of non-genetically modified (GM) crops with GM material, and corporate concentration in the biotechnology industry. Over the course of my research I found that much of this analysis proved to be accurate, but I also became aware that among producers the decision about whether to adopt GM wheat was not reducible to a hypothetical cost-benefit analysis. In fact, the economic decisions made by farmers hinged on the specificities of local history, cultural practices, and the character of wheat as a biological entity.

Anti-GM movements usually conjure up images of radical activists, wearing biohazard suits or dressed as the grim reaper, slashing and uprooting GM crops in unsuspecting farmers' fields. Indeed, Google image searches of "GM crop protest" and "anti-GM movement" produce a large number of Greenpeace signs, young white activists in European cities, and corn cobs wearing frightening faces. Scattered photos of large peasant protests in India are the only regular interruptions of the image of a European, consumer-centred anti-GM movement. So far, however, there has been very little written about "First World" farmers as subjects and leaders of movements against genetic modification. This book approaches the problem of genetic modification from the perspective of "First World" producers (farmers), asking: How have farmers framed their struggle against GM wheat? How have farmers engaged with consumer-centred discourses about genetic modification? To what extent are farmers concerned with the profits produced through farm labour being siphoned off by large agricultural corporations?

Unlike the consumer-driven anti-GM movements that have received much academic and popular attention,[4] the coalition at the centre of this movement was composed primarily of rural and agricultural groups (six of the nine groups represented farmers or rural interests; see Table 1 for a list of the organizations and their positions on GM wheat). In Canada, producer-led activism was key in keeping RR wheat at bay, especially in the provinces where wheat has dominated the agricultural landscape, both historically as a frontier crop and today as an important part of most farmers' rotations. In these provinces, especially Manitoba and Saskatchewan, social struggles over the organization of the farm sector and the corporate extraction of surplus (production over and above what is necessary to reproduce the farm and the farmers) from agriculture have been central issues since the settling of the land in the late 1800s.

More than fifteen years after its introduction, there is still widespread disagreement among academics, activists, scientists, regulators, and others about how best to understand genetic modification, what the possible negative and positive effects of the technology are and will be, and with which arguments to formulate the best critiques. Variously positioned social-movement groups in a wide diversity of geographical locations have emphasized concerns about the moral implications of "playing god" or patenting life forms, the safety of foods derived from genetic modification, the ecological impacts of introducing novel plants into the environment, the loss of farmers' ability to save seed, the colonial practice of "biopiracy,"[5] and the control that biotech companies are gaining over agriculture, science, and regulatory apparatuses. Despite this extremely wide-ranging set of issues, public and academic debate seems to coalesce around either pro- or anti-GM discourses. In investigating the politics of opposition to GM wheat in Canada it became clear to me that these abstracted narratives that posit genetic modification as either bad or good simply do not apply. Many of the Canadian organizations that resisted GM wheat were not actually against genetically modified organisms (GMOs). They were opposed to a particular modification in wheat (and not in other crops) because of the specific agronomic and economic challenges it might pose. In fact, this research questions the possibility of a singular and coherent global movement against GMOs. If future opposition movements want the support of producers, they will have to take into account the particularities of how farmers earn their livelihoods and the specificities of biological organisms.

The wide variety of concerns and interest groups represented in movements against genetic modification can be understood as both a strength and a challenge for a more global opposition to GMOs. On the one hand, diversity provides multiple points for attack and engages many perspectives and people in the movement. In the case of the Canadian campaign against GM wheat, the coalition credits part of its success to the unusual alliance between farm organizations and environmental organizations like Greenpeace. Because of the expertise that each organization had already established in its particular field, the coalition was able to criticize the introduction of GM wheat on multiple grounds, making it more difficult for regulators and politicians to discredit the coalition. On the other hand, the plethora of concerns mobilized by movements against genetic modification make it complicated to carve out space for a nuanced opposition or to link the discourse of opposition to GMOs with other global agricultural movements. For example, organizations that believe the problem with genetic modification is corporate control might find greater affinity with movements challenging global capital than with other anti-GM campaigns. It is not at all clear how organizations with categorical fears about Frankenfoods can engage in a common struggle with groups that want to harness the potential of genetic modification for a more socially and ecologically just agriculture through public, accountable institutions.

While much of the story of GM wheat in Canada is rooted in local histories and ecologies, farmers around the world are dealing with similar processes and threats from global capital, represented by corporations like Monsanto. An understanding of the specific struggle of producers in Canada can be useful for others. But the real moral of this story is that campaigns against genetic modification that want the support of farmers will need to bridge global and local concerns in order to be successful. The contingencies associated with individual crops, local histories and cultures of livelihood-making, and the multiplicity of arguments and organizations against genetic modification mean that temporary anti-GM coalitions are likely to continue to spring up around particular modifications in specific locales. These coalitions will need to rely on the ongoing work of organizations and movements on issues such as the increasing corporate control of agriculture and the movement away from publicly funded, accountable agricultural research. While anti-GM campaigns will necessarily be specific and local, they will also need to address these broader concerns in order to secure long-term viability and effective opposition.

Setting THE STAGE

This book is about farmer resistance to one particular genetic modification (Roundup resistance) in one particular crop (wheat). It is, therefore, essential to provide some background on Monsanto and its RR crops and to show the importance of wheat production and marketing to prairie agriculture. The farmers involved in the coalition to stop RR wheat came from various different organizations and worked alongside health and environmental groups. Each one of these organizations has a rich and complex history and politics that I can only briefly touch on in this book, but I do my best below to provide the reader with some context about the different organizations and their positions on RR wheat. In this opening chapter I also provide the reader with some details about my research methodology and lay out the theoretical concepts I believe are essential to understanding this book.

MONSANTO'S ROUNDUP READY

Roundup is the brand name of a broad-spectrum herbicide produced by Monsanto since 1974. Its active ingredient, glyphosate, is effective against a range of broadleaf and grass plants, including perennial weeds. Monsanto has always faced competition for this herbicide from companies like Bayer Crop Science, which produces a broad-spectrum herbicide called Liberty. Anticipating that the 2001 expiration of its Canadian patent on Roundup would bring greater competition, Monsanto began to experiment with genetic engineering. The company isolated a gene that is resistant to Roundup and began genetically engineering this gene into corn, cotton, soy, and canola seed. The first Roundup Ready crops became available for sale to North American farmers in the early 1990s. Because these crops are genetically engineered to be resistant to the Roundup herbicide, Monsanto now holds a new patent for the gene rather than the herbicide. A farmer must pay a per-acre fee and enter a legally binding technology use agreement in order to purchase and use Roundup Ready seed. Among other provisions, this controversial agreement prevents the farmer from saving seed for subsequent years and gives Monsanto licence to inspect and copy farm records and documents.

Monsanto is an American-based multinational corporation founded by John Queeny in 1901 in St. Louis, Missouri. The company produced saccharin (an artificial sweetener) for Coca-Cola and soon expanded to provide other industrial chemicals to various companies. By the 1960s, Monsanto was manufacturing several of its own widely-used and toxic chemicals, including polychlorinated biphenyls (PCBs) and Agent Orange, a defoliant used by the U.S. army in the Vietnam War. Although Monsanto was not the only producer of Agent Orange, its solution was found to have higher concentrations of dioxins than others, resulting in more devastating effects on humans.[1] Monsanto also became an important promoter of plastics, Astroturf being its most successful product. In the 1980s, Monsanto started to shift its focus to biotechnologies and invested heavily in biotech research and development, becoming one of the first companies to do so. Around this time it bought up several seed companies from around the world. In 1996, the corporation split in two, spinning off its chemical operations into a new company called Solutia (which carries all liabilities associated with Monsanto's industrial chemical past) and focussing its remaining operations around the life-science industry. It now boasts of being an agricultural company "invest[ing] almost $1.5 million a day to look for and bring to market the innovative technologies that our customers tell us make a difference on their farms."[2] The company now claims to be focussed in four main areas: genomics, biotech transformation, seed, and chemistry (centred on Roundup).

WHEAT PRODUCTION AND MARKETS

The possibility that Canada's wheat markets could be jeopardized by the introduction of GM wheat in Canada became a crucial argument for the coalition against RR wheat. It is therefore important to present some data on the significance of wheat exports and the major importers of Canadian wheat. Much of the data below is focussed around the year 2001 to give a sense of wheat production and markets around the time that the coalition against GM wheat announced its opposition in July 2001. In Canada, the greatest area planted to wheat is found on the Canadian prairies, with smaller concentrations in southern and eastern Ontario (Figure 1). The arid land of the prairie provinces produces lower yields but higher protein content, a quality characteristic that is highly valued by flour millers.[3] This inverse relationship between aridity and protein content creates the perplexing statistic that the provinces that grow the most wheat have the worst yield ratios (Figure 2).

The relatively high protein content of prairie wheat has been carefully marketed by the Canadian Wheat Board, which, until August 2012, was the monopoly marketing agency for the export of wheat and for domestic human consumption in the three prairie provinces and a small portion of British Columbia. While the rest of the data below is given for Canadian wheat, it provides a good representation of prairie wheat realities since the bulk of wheat in Canada is produced in the prairie provinces.

Figure 1. Wheat area distribution in Canada

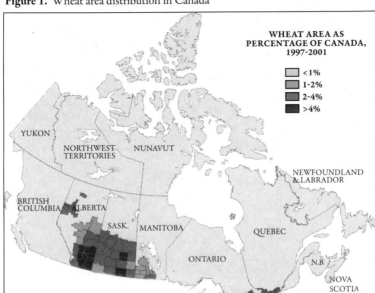

Source: United States Department of Agriculture, Production Estimates and Crop Assessment Division of the Foreign Agricultural Service, with information based on data from the agricultural division of Statistics Canada, http://www.fas.usda.gov/remote/Canada/can_wha.htm.

Canada is one of the world's major producers, producing roughly 4 percent of the world's wheat (Figure 3). This is significantly less than what is produced by the United States (roughly 10 percent) and by the EU as a whole (roughly 21 percent), but more than other major wheat producers like Argentina (2 percent) and Australia (3 percent). More significant is Canada's 15 percent share of the world wheat trade (Figure 4). It is clear that wheat is important for Canada as an export crop; relatively little of its total production is reserved for domestic consumption.

Figure 2. Average wheat production by province, 1997-2001

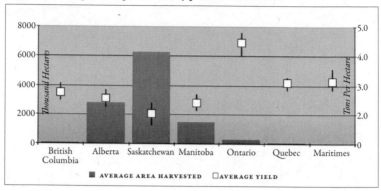

AVERAGE AREA HARVESTED   AVERAGE YIELD

Source: United States Department of Agriculture, Production Estimates and Crop Assessment Division of the Foreign Agricultural Service, with information based on data from the agricultural division of Statistics Canada, http://www.fas.usda.gov/remote/Canada/can_wha.htm.

Figure 3. Major wheat producers

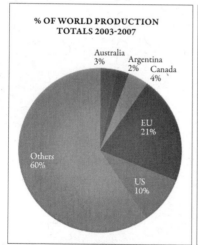

% OF WORLD PRODUCTION TOTALS 2003-2007

Australia 3%
Argentina 2%
Canada 4%
EU 21%
US 10%
Others 60%

Figure 4. World wheat trade

MARKET SHARE JULY/JUNE 2003-07

Australia 12%
Argentina 9%
Canada 15%
US 26%
Others 26%

Source: Canadian Wheat Board, 2007–08 Annual Report, http://www.cwb.ca/public/en/about/investor/annual/pdf/07-08/2007-08_annual-report.pdf(accessed October 2008).

As of the 2000–2001 crop year, and over the ten years ending in 2001, the five top importers of Canadian wheat were Japan, the U.S., Mexico, Iran, and China (Table 2). Significant to the story of GM wheat was Japan's importance as an importer of Canadian wheat. Losing Japan as an export market for wheat because of its rejection of genetic modification could have been devastating for the wheat industry. Western Europe was another region that threatened to reject GM wheat, and while it accounted for only 8.8 percent of Canadian wheat exports in the 2001–2002 crop year (Table 3), it tended to buy high-quality, high-priced wheat and had been a key importing region since Canada began exporting wheat in the mid-nineteenth century. Figure 5 gives the total monetary value of Canadian wheat exports from 1998 to 2008, averaging $3.84 billion over that period.

METHODOLOGICAL APPROACHES

I began my research on the politics of GM wheat on the Canadian prairies more than two full years after Monsanto had discontinued its breeding and field research on Roundup Ready wheat in Canada and the United States. Having followed the politics around RR wheat's introduction through the media for the preceding five years and having read what academics had said about Canadian biotech policy and regulation, I had not thought twice about using words like "politics" and "struggle" when presenting my research interests to potential participants. The initial interviewees for this research embraced these terms and set out to convince me of their positions of opposition. It seemed natural to begin my interviews with representatives from the organizations that were formal members of the coalition against RR wheat that announced itself at a press conference in Winnipeg in July of 2001. From these interviews, and from an examination of all articles pertaining to genetic modification in Western Canada's most prominent weekly farm newspaper from 2000 to 2006, I learned that a number of other actors were essential to the story of RR wheat. I expanded my field of research to include inter-

Table 1. Organizations involved in the 31 July 2001 coalition to stop the introduction of Roundup Ready (RR) wheat

NAME OF ORGANIZATION/ DATE OF FOUNDING	TYPE OF LOBBY	MAIN COMPLAINT(S) ABOUT RR WHEAT	PROPOSED ACTION
National Farmers Union (NFU)/1969	Left-wing farm organization formed to unite provincial farmers' unions that had led radical farm organizing since World War I	Loss of control of the food/seed system to multinational corporations, threat to profitability and autonomy of family farm	Moratorium on all GMOs. All GMOs must be subject to democratic control, collective ownership, and not-for-profit distribution
Saskatchewan Association of Rural Municipalities (SARM)/1905	Advocate of rural municipalities to senior levels of government	Loss of markets, secrecy of field trial locations	Ban GM wheat until segregation and detection systems, tolerance levels, markets, and changes to regulatory system are established
Saskatchewan Organic Directorate (SOD)/1998	Producer-controlled umbrella organization for producers, processors, buyers, traders, certifiers, and consumers	Liability in cases of contamination, and loss of ability to farm organically	Complete ban on all GMOs since contamination is inevitable
Agricultural Producers Association of Saskatchewan (APAS)/1999	Saskatchewan general farm organization with representation from all rural municipalities	Market impact, agronomic issues, effects on zero till	All varieties of GM wheat must be approved based on agronomic merit, including market acceptability
Keystone Agricultural Producers (KAP)/1984	Manitoba general farm organization	Market impact, agronomic issues, the impossibility of segregation from non-GM crops	Prevent registration until consumer acceptance
Canadian Wheat Board (CWB)/1935	Western Canadian single-desk marketing organization jointly governed by producers and the federal government	Loss of markets (more than eighty percent of customers are concerned about GM wheat).	Add cost/benefit analysis to regulations. Do not release RR wheat at this time.
Canadian Health Coalition (CHC)/1979	Non-governmental organization (NGO) primarily concerned with public health care	GMOs may have negative health impacts; regulatory system is anti-democratic and serves life-science industry	Regulatory system must be overhauled and serve the public
Greenpeace Canada/1971	International environmental NGO that began in Canada	GMOs will harm the environment and may have negative health impacts; life should not be patented	Stop all GMOs, reform the regulatory system
Council of Canadians (CoC)/1985	Multi-issue nationalist NGO	Consumers don't want GM wheat. Long-term impacts on health and the environment are unknown	Stop all GMOs pending labelling, long-term studies, and regulatory reform

Table 2. Major importers of Canadian wheat (thousands of tonnes)

	2000-2001	TEN YEAR AVERAGE (1991-1992 TO 2000-2001)
Japan	1,406	1,346
United States	1,079	1,207
Mexico	1,158	687
Iran	1,532	1,636
China	17	2,643
Ohters	7,875	7,909
Total	13,067	15,427

Source: Compiled from Canadian Grain Commission, *Canadian Grain Exports, Crop Year 2001–2002,* http://www.grainscanada.gc.ca/statistics-statistiques/cge-ecg/annual/exports-2002-eng.pdf.

Table 3. Exports by country of destination for wheat, durum wheat, and wheat flour (2001–2002 crop year)

	TONNES	PERCENTAGE OF TOTAL	BIGGEST RECIPIENT(S) (TONNES)
Western Europe	1,378,399	8.8	Italy (562,261) UK (361,301)
Eastern Europe	33342	0.21	Poland (33,342)
Africa	3,013,496	19.25	Algeria (800,158)
Asia	5,484,013	35.02	Japan (1,370,636)
Western Hemisphere	5,706,118	36.44	USA (1,899,084)
Oceania	45,459	0.29	New Zealand (45,459)

Source: Compiled from Canadian Grain Commission, *Canadian Grain Exports, Crop Year 2001–2002,* http://www.grainscanada.gc.ca/statistics-statistiques/cge-ecg/annual/exports-2002-eng.pdf.

Figure 5. Monetary value of total Canadian wheat exports, 1999–2008

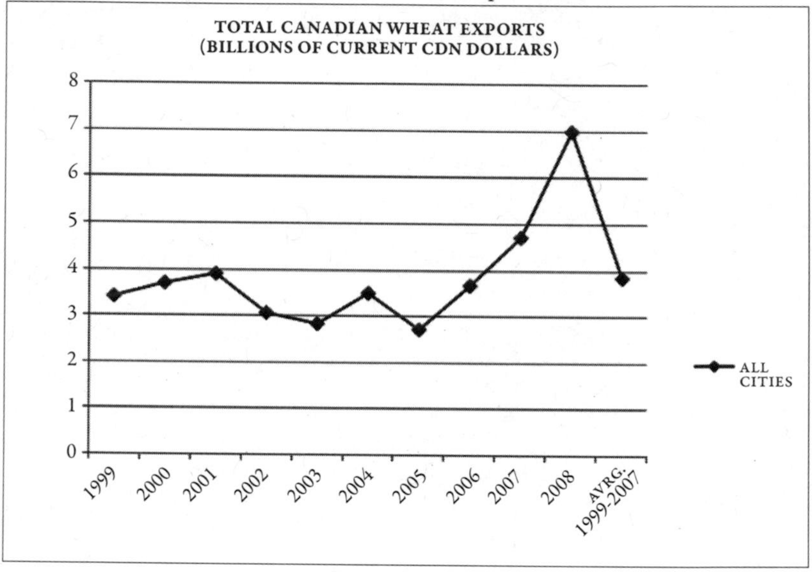

TOTAL CANADIAN WHEAT EXPORTS
(BILLIONS OF CURRENT CDN DOLLARS)

Source: Industry Canada, Trade Data Online Database, "Trade by Product", http://www.ic.gc.ca/sc_mrkti/tdst/tdo/tdo.php#tag.

views with actors from all sides of the debate, including plant breeders, scientists, biotech lobby groups, industry organizations, the Canadian Biotechnology Advisory Committee, a representative from Monsanto, a representative from Saskatchewan Agriculture and Food, and farm organizations that publically supported the introduction of RR wheat.

The farm organizations involved in the coalition against RR wheat comprised the vast majority of farm groups on the prairies. The Wild Rose Agricultural Producers (WRAP), the general farm organization from Alberta, was notably absent from the coalition, since its equivalents in the two other prairie provinces were quite actively involved.[4] The Saskatchewan general farm organization, the Agricultural Producers Association of Saskatchewan (APAS), is a newer organization (founded in 1999) and has been somewhat less stable than its Manitoban counterpart, the Keystone Agricultural Producers (KAP). While membership figures for farm organizations are generally not available and not released to the public, these farm organizations likely represent the largest number of farmers in their provinces. The National Farmers Union (NFU) represents fewer farmers, but has been a long-standing and outspoken feature of agricultural politics since its founding in 1969, and before that through its predecessors—first the United Farmers of Canada and then provincial farmers' unions. Thus, the NFU has roots in the radical populist organizing of the first half of the twentieth century that resulted in cooperative grain elevators, political alliances with labour, and pools for wheat. The Canadian Wheat Board (CWB), which played a central role in the coalition, was one of the products of this early organizing and was supported by the vast majority of the farming population. In recent years, it has been marred by controversy over its monopoly and governance and was eventually dismantled on 1 August 2012 through an act of Parliament.[5] Finally, the Saskatchewan Organic Directorate (SOD) is a relatively new group (founded in 1998) that has an increasing membership due to the growth of organic farming in Saskatchewan. Despite representing a minority of producers, it has received much media attention (because of its pursuit of a class action against Monsanto and Bayer for the contamination caused by GM canola) and recognition and support from the provincial government.

As will become clearer in Chapter 4, the coalition struggled with its internal coherence. Not only were farm and rural groups reluctant to work alongside urban environmental and health groups like Greenpeace and the Council of Canadians, but the great diversity of farm organizations involved in the coalition also posed a challenge for organizing. More mainstream farm and rural organizations like APAS, KAP, and the Saskatchewan Association of Rural Municipalities (SARM) understood themselves as representing the vast majority of conventional prairie farmers, while the SOD and the NFU were often painted as representing non-conventional farmers and taking more marginal and radical positions against GMOs. Despite the discordant reasons for their opposition (see Table 1), there was a high degree of consensus on a few key issues that are taken up most centrally in this study.

Of course organizations were also internally fractured. The SOD in particular continues to struggle with its connection to the wider organic movement that espouses principles such as local food, small-scale intensive agriculture, and mixed farming. Its membership includes both producers who strive to incorporate these wider principles and farmers who are mainly driven by niche markets, who perform to the minimum organic standards and are not particularly committed to the philosophy of the movement. Representing the largest number of farmers, organizations such as KAP and APAS have tended to focus on the issues that are least divisive, for example, securing better support programs for farmers from provincial and federal governments. While internal fissures and disagreements must have certainly played a part in the negotiated positions reached by the organizations and the coalition, these were not subjects that my participants wanted to discuss. My observations at farm meetings provided a certain level of insight into such cleavages, but this discordance was largely beyond my access given my position as an academic outsider. Moreover, the organizations involved in the coalition had strategic reasons to present their opposition as consensual so as to maintain strong public resistance to RR wheat. While I do not wish to underplay the contentious nature of intra-organizational politics, all the representatives I interviewed stressed the overwhelming support among the members of their organizations for their opposition to RR wheat.

Some commodity organizations took vocal positions of support for RR wheat, and these tended to be organizations with corporate involvement. Of those supporting RR wheat, the Canola Council of Canada likely represents the largest number of farmers. This organization draws its membership not only from producers but also from processors, agro-chemical and seed companies, and other corporate actors. The Canola

Council supported RR wheat because of the commercial success of GM canola varieties. The Western Barley Growers Association was also a very vocal supporter of RR wheat; it is comprised of producers, industry, and end-users. This organization and the Western Canadian Wheat Growers Association have pushed for uninhibited functioning of free markets in all areas of Western Canadian agriculture. Membership figures for these organizations are not available, but the Western Canadian Wheat Growers Association is not thought to represent a very large number of farmers, especially given that it disbanded for a short period and re-established its operations in early 2004. The Grain Growers of Canada represents a number of these commodity associations at the national level and is also involved in fighting for market liberalization.

. . . .

As a method for conducting social science research, interviews are commonly justified by the epistemological assertion that humans are competent reporters of both their past and present attitudes, beliefs, behaviours, relationships, and interactions[6] However, interview subjects are influenced in all sorts of ways that affect how they interpret and give accounts of their experiences to a probing outsider. In fact, common caution given by introductory research methods texts is that respondents cannot always be trusted to tell the "truth, the whole truth, and nothing but the truth" about their experiences.[7] This type of concern reflects a positivist understanding of reality where there are believed to be objective truths about what people think and do and how they interact. For positivists, the collection of data through interviews can be problematic as it is difficult to know which statements made by an interviewee are true, and which are manipulations of reality. Respondents may have cultural or strategic reasons for presenting information about themselves and others in a particular light, and may leave out or distort some information.

Another epistemological critique of the interview as a method has been mounted by postmodernists, who have no faith that interviews reveal the "truth" about the actions, transactions, and beliefs of respondents. Postmodernists contend that interviews tell the researcher nothing beyond the accounts that people give.[8] From this position, interviews are a topic *for* social research, rather than a method of conducting social research. A postmodernist may engage in an analysis of an interview as text or discourse, but would disagree that that discourse is representative of anything (idea, practice, or belief) beyond itself.[9] For a postmodernist, the interview is not necessarily an effective means to gain insight into people's beliefs and motivations since it tells us little beyond that encounter.[10]

I understand the interview as both a text that can be read as a topic *of* research and as a vehicle for revealing and generating knowledge beyond the text. I think these two epistemological positions must not be mutually exclusive. Similar to the postmodernist, I am interested in the discourses of my participants and the ways in which they have narrated accounts of their experiences. Unlike the postmodernist, I am taking the position that a link can be made between what a participant says they do and think and what they do and think in practice (or in situations and spaces beyond our interaction). This link, rather than being an objective "truth," is interpreted and presented through a subjectivity (the interviewee) and received and (re)interpreted through a second subjectivity (the researcher). This is the sense in which the interview is an intersubjective production where meaning is not objective, but rather an ongoing interpretive accomplishment.[11] It is not my intention to be suspicious of my participants' motivations and accounts, but rather to pay special attention to how they have presented their opinions and practices to me, a subject situated differently from others with whom they may interact. I am concerned not that my respondents have left out or manipulated aspects of their stories, but rather with why they told me their stories in certain ways, and how they represented their practices to me.

I take the position that not all explanations are equally compelling and that the researcher has a duty to differentiate among them with the knowledge that a single coherent truth is unattainable. For example, when I heard a similar narrative from a wide variety of participants while conducting interviews, I knew that it was likely to have been important to the way that the politics of RR wheat played out. Whether or not it was a "true" reflection of reality was less significant than the fact that it was doing real work for my participants. However, a researcher must not blindly accept all prominent explanations. I also used a number of written documents, including press releases, policy statements, and farm newspaper articles, in order to gauge the importance of particular interpretations and explanations. Here again, it was important to consider these texts as authored by particularly positioned individuals who are part of larger webs of power. In order to differentiate among possible explanations, I used a combination of criteria including the frequency of the account, the extent to which it was reproduced across differently positioned actors and data sets, and explanations found in academic bodies of knowledge. I used press releases and policy documents to characterize organizations' public positions. These written documents were usually the outcome of a collaborative effort and were artifacts of particular moments in time. Since almost all of my interviews were one-on-one interactions, it was important to consult such written documents to establish the negotiated public positions of the collectives.

I performed semi-structured, in-depth interviews with forty-three participants and loosely followed an interview guide—the questions I asked of participants changed significantly based on their involvement in the public debate. I always interviewed participants as representatives of their organizations, yet participants often used their personal experiences to answer my questions. Farmers often represented their experiences as common to farmers more generally. Interviews varied in length, but most were between one and one-and-a-half hours.

Another important source for this research was Western Canada's largest weekly farm newspaper, the *Western Producer*. I examined every article pertaining to genetic modification between the years 2000 and 2006. This source allowed me to piece together important events in the debate over GM wheat, gauge the weight of the different arguments for and against GM wheat, and follow up on the accounts of my interview participants where needed. I used the *Western Producer* at the beginning of my research to familiarize myself with important players and debates and at the end to compare what I was finding in my interviews with journalistic accounts.

Finally, I observed at five farm meetings. Although none of these meetings were directly on the topic of genetic modification, each meeting gave me a sense of some of the long-lasting effects of Monsanto's decision to abandon GM wheat. At these meetings it was patently obvious that the issue of genetic modification had not left farmers' minds and that they were carefully considering how, for example, proposed changes to the regulation of grain in Canada would affect the introduction of new GMOs. The information I gleaned from these meetings served mostly to provide context.

THEORETICAL APPROACHES
This book fits within a larger body of literature on agrarian questions. That is, it is concerned with some classical questions about the relationship of capitalism to agriculture, and about the nature of farmers and peasants as historical subjects. With regard to the first question, there are two opposing views: those who emphasize the continual expansion of capital relations into and through agriculture, and those who insist on the unique qualities of agricultural production that make it resistant to complete absorption by capitalism.

Writing *The Development of Capitalism in Russia* in 1899, Vladimir Lenin was at pains to show how the capitalization of the countryside created two new classes of rural inhabitants, obliterating the distinct peasant class through a process of differentiation.[12] The first new class, the rural bourgeoisie, is comprised of independent commercial farmers (who employ wage workers and perform little farm work themselves) and owners of commercial and industrial establishments (who combine commercial agriculture with industrial activity). The second is a rural proletariat composed of poor peasants—primarily allotment-holding labourers of all types, but also landless workers. According to Lenin, the process of differentiation that resulted in the formation of the two classes would drive towards completion, obliterating the middle class of subsistence-producing peasants and laying the ground for socialist revolution.

Also stressing capitalist transformations in the countryside, Karl Kautsky's *Agrarian Question,* written in 1899, contemplates the best agricultural program for the Social Democratic Party of Germany.[13] He argues that the proletarianization of the peasantry should not be celebrated, but nor should it be impeded from progressing. At the very least, proletarianization freed peasants from the land, and the consolidation of large capitalist farms led to impressive productivity. Unlike Lenin, however, Kautsky identifies a number of characteristics of peasant farming and peasant subjectivity that made the development of agrarian capitalism tenuous. For example, he recognizes the tendency among small farmers to overwork and underconsume, stating that "small farms have two major weapons to set against the large. Firstly, the greater industriousness and care of their cultivators, who in contrast to wage labourers work for themselves. And secondly, the frugality of the small independent peasant, greater even than that of the agricultural labourer."[14] Furthermore, Kautsky points to the greater difficulty of employing machines in agriculture versus in industry because of the unevenness of nature (for example, the uneven terrain and the seasonal nature of production). Finally, Kautsky characterizes the small landholder as inhabited by "two souls": the peasant and the proletarian. Conservative parties have an interest in protecting the peasant "soul" who is the owner of his/ her means of production and has the capacity for entrepreneurialism. Social democrats, Kautsky argues, should support the proletarian "soul," whose interests are the social development of society.[15]

Writing against Lenin, Alexander Chayanov insisted that the middle peasantry was not a disappearing class in Russia and that peasants would continue to persist because of the unique logic of peasant economy.[16] Whereas Lenin estimated the "middle peasantry" to comprise around 30 percent of the rural population in the late 1800s, in 1925 Chayanov estimated that "ninety percent of the total mass of peasant farms are pure family farms."[17] These were farms that relied only on family labour, even if their members were also engaged in handicrafts and commodity production. According to Chayanov, the family-labour farm organized its production so that equilibrium was reached between the drudgery of labour and the satisfaction of the consumptive needs of the family. It follows that peasant families had been able to resist differentiation precisely because they worked longer and harder and paid more for certain factors of production (for example, for the renting and buying of land) than would capitalist firms/farms. It is not that social differentiation was absent in the Russian countryside, but Chayanov explained much of it as a product of the cycles of demographic change within families. Farms expanded and bought up more land when they were large and comprised of dependent children; when children moved away they contracted.

Chayanov's insights about the particularities of peasant economy were taken up with new fervour in the last decades of the twentieth century. Harriet Friedmann, for example, showed that the differentiation and capitalization of the countryside identified by Lenin and Kautsky as inevitable had not taken place in all parts of the world.[18] In fact, the frontier wheat economies of the United States and Canada, based on family units of production, were outperforming capitalist wheat farms in Europe by the time that Lenin and Kautsky were writing. During the early 1900s, household production of wheat continued to grow, and by 1935 production based on family labour accounted for the vast majority of wheat on international markets. Friedmann explains the outperformance of capitalist farms by household production as a product of the different requirements of the two forms of production. Drawing on Chayanov, she shows that the net product of household production is structurally identical with the fund for familial consumption and reproduction—there is no necessity for a surplus product like there is with capitalist production. Thus, households make their production decisions based on family needs and desires. Some members may also engage in wage labour to ensure the reproduction of the farm. The emphasis in this work is on the capacity for family-farm production to resist both destruction and absorption by capital.

On top of the household organization of production, Susan Mann and James Dickinson have identified nature as an obstacle to the development of capitalist agriculture.[19] Taking their inspiration from Marx's discussions of labour time, production time, and the time during which agricultural products grow, Mann and Dickinson advance the thesis that capitalism develops in those spheres in which the gap between production time and labour time can be reduced. Nature presents certain barriers to the development of agrarian capitalism, the biggest of which being the period after planting and before harvest when the crop is maturing through natural processes and labour is relatively idle. Whereas nature is an input in industry, in agriculture it acts as the factory itself. In the Mann-Dickinson thesis the agrarian question takes a geographical turn. The materiality of nature takes centre stage and crops are distinguished from one another based on their turnover times, their seasonalities, and the degree of unity in their production and labour times. In a subsequent work that reaffirms the Mann-Dickinson thesis, Mann shows how capitalist development varies by commodity.[20] Extensive crops, for example, tend to have a higher composition of capital (that is they employ more non-living inputs like machinery) and to be organized as family production units rather than capitalist farms. Furthermore, crops that take longer to mature are more likely to resist capitalization.

The Mann-Dickinson thesis, where nature acts as an obstacle to capital, has been met with much criticism. For example, David Goodman, Bernardo Sorj, and John Wilkinson emphasize the capacity of industrial capital to get around the "problem of nature" through practices of appropriationism and substitutionism. Appropriationism is the processes of industrial appropriation of activities related to farm production and processing; industrial capital expands into and takes over rural activities and labour processes.[21] Initial examples of appropriationism were in the realm of farm labour (mechanization), then in the chemical properties of the soil (fertilizers), and next in the actual production process itself (hybrid seed). Here, elements of the production process that used to be firmly within the control and economy of family labour get siphoned off by industrial capital. Substitutionism is a set of processes that replace agricultural products with industrial substitutes, thereby eliminating the rural production process in that commodity.[22] Examples include the industrial manufacturing of powdered milk, aspartame, margarine, dyes, and rayon. For these authors, agriculture is diminished to the residual activities that have resisted transformation by industrial capital, and substitutionism and appropriationism will continue to chip away at what is left. Critical of the processes of substitutionism and appropriationism, Goodman, Sorj, and Wilkinson, argue that they result in the "freeing" of production from land and natural processes and the reduction of agriculture to the production of raw materials and biomass for industrial processing.

In his study of the commodification of seed through hybridization and genetic modification, Jack Kloppenburg has also emphasized the ways in which obstacles posed by nature have been overcome by industry.[23] In order to successfully wrestle the reproduction of seed away from public institutions and individual farmers, Kloppenburg argues that two simultaneous routes needed to be pursued. First, a technical strategy was developed wherein the scientific practices of breeding and later genetic manipulation were used to strip seeds of their natural properties of reproduction, and thus make them more amenable to commodification. Second, industry pursued changes in legislation that would allow the expansion of private property, thus negating the legal right of farmers to harness the reproducibility of seed. In the story of plant breeding in the United States (and indeed all over the world) capital relations have thoroughly entered into and transformed agriculture without totally displacing the class of non-wage labourers whose production is based in the family.

George Henderson agrees in his history of California that capital has found ways to get around the obstacles that nature presents for agriculture, but he also argues that these same obstacles confronting industrial capital paradoxically serve as opportunities for the circulation of financial capital.[24] The disunities of production and working time and the seasonality of crops elaborated by Mann and Dickinson result in the need among farmers for loans and credit to cover the periods of idleness and the gap between planting and harvest. Credit is, thus, a method of extracting value from agriculture without, necessarily, the development of wage labour and capitalist farms. In fact, Henderson argues that in the late-nineteenth and early-twentieth centuries agriculture became an outlet for the capital that had already amassed in Californian cities through rapid urbanization and a boom in mining. The flow of financial capital into agriculture helped enable the transition to intensive agriculture through investment in irrigation and through land speculation. The increase in the price of land meant that farmers of all kinds had to take out large mortgages, further perpetuating the extraction of surplus by financial capital.

How do the farmers in this research fit into the debates about the agrarian question? Central to this research and to the literature on the agrarian question is an understanding of family farms as productive units employing the majority of their farm labour from within the family. This was the model upon which agriculture was based on the Canadian prairies, but there has been movement away from the family-labour farm as the necessity of economies of scale has grown. Unfortunately, Statistics Canada's definitions and classifications of "census farms" rely exclusively on gross farm receipts and tell the researcher nothing about the organization of farm labour. In the Canadian Census of Agriculture the classification ranges from farms with gross receipts under $25,000 to farms with gross receipts over $1 million. Recent data has shown that from 2001 to 2006 the number of large farms (defined as having over $250,000 in gross annual receipts) has increased 17.8 percent while farms with annual receipts under $250,000 have declined by 10.5 percent.[25] Such data give us a sense that farms are consolidating, but do not provide a good indication of the allocation of labour on Canadian farms.

Regardless of the allocation of labour on prairie farms, it is patently clear that family farms are under assault. Capitalism is penetrating prairie agriculture in an increasingly aggressive manner, and Monsanto's RR wheat can be understood as yet another intrusion. Indeed, Monsanto can be understood as engaging in a process of appropriationism, transforming what used to be a farm practice of seed saving into an industrial and fully capitalist process. This appropriation has relied on legal mechanisms (property rights) that make it illegal for farmers to reproduce Monsanto's seed in their fields. Experimentation with Genetic Use Restriction Technology is also underway as a technical solution that would transform the biology of the plant to make it infertile in subsequent generations so farmers could not reproduce it illegally.

Monsanto's RR wheat is properly understood as a mechanism of privatization and capitalization of Canadian agriculture, yet Canadian farmers successfully resisted this intrusion (at least temporarily) because of their cultural and institutional attachments to wheat and of the nature of wheat as a biological entity. This points to the salience of a commodity-specific approach to the agrarian question, as suggested in the Mann-Dickinson thesis where human-nature relationships present certain obstacles to the intrusion of capital. It also confirms the insights of Kautsky, Chayanov, Friedmann, and others about the very unusual logics that underlie production based on family labour, where decisions about how much to produce, which crops to grow, how much to work, and how many risks to take are based on the needs of the family rather than the imperatives of average profit rates. Prairie farmers fashion their production decisions around a broad range of concerns including the ways in which different plants grow in the fields, the different institutional supports for different crops, family traditions, seed-saving practices, and much more. All of these factors point to the need to understand producer resistance to RR wheat from perspectives other than economistic cost-benefit analyses. While capitalism and capitalists treat all commodities as varying quantities of exchange value (essentially as commensurable to the prices they can fetch in the market), producers understand their labour and its products as embedded in moral and cultural economies that are reflective of the heterogeneity that makes up human and non-human life.

This kind of vibrant diversity would be echoed in the wide-ranging coalition that came together in 2001 to contest the introduction of genetically modified wheat in Canada.

Regulating
AND PROMOTING
Biotechnology in Canada

CHAPTER TWO

In the years leading up to the commercialization of the first products of biotechnology in the 1990s, governments around the world had to decide how to regulate genetically modified organisms and how to engage, or not, with the biotech industry. Should regulators treat genetically modified plants like their "conventional" equivalents, or should they take a precautionary approach since the full spectrum of possible risks was, and is, as yet unknowable? If risks are to be measured and taken into account, should such an assessment include health, environmental, economic, social, and/or ethical factors? Should the scientific data for such risks be collected by state scientists, independent scientists, or company scientists? If credible risks are identified should the GM plant be abandoned? Should the government invest in and support biotechnology research? Can the state be both an impartial regulator and a supporter of biotech research? In this chapter I address the approach that the Canadian state has taken to regulating the products of biotechnology and engaging with the biotech industry. I also give the reader some background on the process through which GM plants are approved for commercialization in Canada.

CANADA'S BIOTECHNOLOGY STRATEGY

The Canadian state (including political, bureaucratic, and public institutional branches) has been subject to intense criticism by academics and social movements opposing genetic modification. Indeed, early promotion of biotech research and development in the 1980s conditioned the Canadian government's implementation of its regulatory regime and shaped its engagement with the public's concerns related to health, environment, and economy. The result has been an aggressive promotional and top-down approach to biotechnology with regulatory agencies only belatedly and superficially engaging with public concerns. This approach to biotechnology research, investment, and regulation in Canada contrasts with the more precautionary stance taken by many EU countries and the more participatory approach taken by countries like New Zealand.[1]

Canada's first national biotech strategy dates back to the early 1980s, when a series of task forces and reports were produced by and for the federal Ministry of State for Science and Technology. According to Elisabeth Abergel and Katherine Barrett's thorough chronological analysis of Canadian biotechnology policy, the goals of the strategy were, from the beginning, to aggressively pursue commercial progress in biotechnology.[2] Amid economic recession and lagging manufacturing industries, biotechnology was understood as a strategic technology that could foster economic growth in a new era of competitive innovation in high-technology industries. During this initial engagement with biotechnology, the Canadian government encouraged commercial progress by increasing the role of private companies in research and commercialization through financial investment and incentives like tax shelters. The focus of Agriculture Canada's research was also brought in line with the goals of biotechnology development and commercialization. In 1983, the federal government gave $22 million over two years to fund Canada's first National Biotechnology Strategy and over $100 million to fund national biotechnology research centres, with the Saskatoon centre playing a leading role. It was also well understood that in order to gain a competitive advantage vis-à-vis the rest of the world, legislation and regulations would need to be reconfigured to meet the needs of the biotech industry and the research community.[3] In short, the Canadian government fostered an industry-friendly environment with considerable state support in order to achieve Canada's goal of becoming one of the world's leading producers of biotechnology.

In 1987 Canada's commitment to the high-tech industry, with strategic focus on biotechnologies, was reinforced through a broad new Science and Technology policy called InnovAction. According to the National Advisory Board on Science and Technology the new "S&T [Science and Technology] policy should enhance, not inhibit, the speed and intensity of innovation to support economic growth, improvements to quality of life, and the advancement of knowledge. It should strengthen the capacity of all sectors of the economy to use the results of S&T. Key activities include legislation and policy development (e.g., regulation, standard setting, intellectual property rights, and environmental protection), as well as promotion of international, regional and sectoral coordination, and the encouragement of technology transfer and diffusion."[4] Alongside policy support, the government also committed a large percentage of its science and technology expenditures to agricultural biotechnology. From 1998–2002, for example, the Canadian government spent $56 million per year on biotechnology science and technology—16 percent of total science and technology expenditures devoted to agriculture.[5]

The Canadian state's decision to aggressively pursue the development of biotechnology should be understood in the context of what sociologist Peter Andrée has identified as a transnational biotech bloc. This bloc "is spearheaded by a handful of agrichemical [and pharmaceutical] corporations, but...also involves promotional and regulatory arms of government and civil society institutions (such as universities), that have worked together to realize a particular vision of genetic engineering in agriculture."[6] Andrée argues that the transnational biotech bloc has successfully (though unevenly across space) promoted its own interests in the development of biotech industries as the common interests of society as a whole. It has done this through discursive, material, and organizational manoeuvres: through public relations campaigns; by promoting private property regimes in plants and novel technologies (so as to secure biotech profits); and by winning over public scientists, universities, and regulatory structures. Importantly, the transnational nature of the biotech bloc heightens the perception of competition between states and promotes the perceived benefits of "early adoption."

Buying into the rhetoric and material promises of the transnational biotech bloc, Canada has promoted biotechnology from the beginning as a strategy for economic development and international competitiveness. Regulations, intellectual property rights, and international trade have been fashioned around the needs of the biotech industry, with private interests increasingly determining the nature of the Canadian state's engagement with the technology. Indeed, the development of the biotech industry in Canada happened alongside a policy change that saw federal agricultural research move from relative autonomy toward increasing engagement with public-private partnerships and industry-driven research agendas and funding.[7] Before consulting the public or developing a regulatory structure for the approval of GMOs, the Canadian state had already enrolled in the transnational biotech bloc.

REGULATING THE PUBLIC

Canada's early promotion of biotechnology set the course for its regulatory policy and for the state's engagement with public concerns around the technology. Two concurrent developments posed challenges for the Canadian state's promotional approach to biotechnology. First, in order to commercialize the first products of biotechnology a regulatory framework was needed to govern field tests and the safety of the novel plants for the environment and public health. This would put the government in the uncomfortable position of being both regulator and promoter of biotechnology. Second, starting in the late 1980s a growing number of activists, NGOs, and scientists began voicing concerns about the potential ethical, environmental, social, and health risks associated with GMOs. In response, the state would need to downplay its promotional role in order to gain the trust of its citizens as a credible guardian of public interest.

Amid concerns around the risks associated with GMOs, the government of Canada began to cobble together a regulatory framework. From the beginning the government's approach has been to give no special treatment to products developed through the use of genetic engineering. In this way, the Canadian state, unlike European states, has refused to regulate based on genetic engineering as a novel *process*. Instead, the state has insisted on using existing legislation on environmental and health safety and making regulatory decisions based on the novelty of the *product*, whether derived through genetic engineering or not. Genetic engineering is here understood as the latest technology in a long trajectory of plant breeding, so that food derived from genetic engineering is substantially equivalent to (i.e., of no more risk than) that derived from any other plant-breeding technique. For example, the first legislation to address biotechnology regulation as a discrete process, the Canadian Environmental Protection Act (CEPA), was passed in 1988. Despite civil-society calls for a comprehensive regulatory framework under the CEPA, the government decided existing legislation in the federal departments of Agriculture and Health would take care of the bulk of the products of biotechnology. CEPA only provided Environment Canada with a mandate to regulate products of biotechnology not already regulated under other legislation.[8] Thus, Agriculture Canada maintained authority over the approval of GM crops through the Seeds

Act, entrenching the idea that products of genetic engineering are no different from those derived through other plant-breeding processes.

Despite the work already being done within Agriculture Canada to establish guidelines and regulations for GM plants and field tests based on the principle of substantial equivalency, the Government of Canada went ahead with a new regulatory framework for biotechnology in 1993. According to Abergel and Barrett, the purpose of the new regulatory framework was to secure public confidence, provide a stable and predictable regulatory climate for industry, and harmonize policy approaches with the U.S. and OECD countries.[9] For example, in a press release, the Government of Canada stated that the goal of the framework was "to minimize environmental risks while fostering competitiveness through timely introduction of biotechnology products to the marketplace."[10] Regulations were, thus, explicitly written to promote the development of more biotech products and to enable these products to reach international markets quickly. The principles of substantial equivalence and product-based regulation were key, as they allowed for fairly uncomplicated environmental assessments for the unconfined release of GMOs. In fact, a new category called "plants with novel traits" was established under which all plants with genetically novel traits, regardless of the process through which they were bred or modified, would be regulated. Under this framework GM plants with familiar traits that are considered substantially equivalent to existing plants need no new safety assessments. The principle of substantial equivalence and the category of plants with novel traits (PNTs) were accepted without challenge by the Canadian Food Inspection Agency when it took over regulatory duties from Agriculture and Agri-Food Canada in 1997.

The regulatory framework that resulted from an understanding of substantial equivalence is summarized on the CFIA's website (Figure 6). For a plant to trigger a regulatory assessment it must be established as possessing a novel trait. According to the CFIA, a trait is considered novel when it is new to stable, cultivated populations of the plant species in Canada *and* has the potential to have an environmental effect. According to the CFIA, to date, all genetically engineered (GE) plants in Canada have been considered to contain novel traits. Once the determination of novelty has been made, CFIA evaluators examine the plant's molecular characteristics, identifying the new or modified genes and establishing how they are likely to behave. This environmental safety assessment examines five categories: (1) the potential of the plant to become a weed or to be invasive of natural habitats, (2) the potential for gene flow to wild relatives, (3) the potential for a plant to become a plant pest, (4) the potential impact of a plant or its gene products on non-target species, and (5) the potential impact on biodiversity. In all cases CFIA evaluators compare PNTs to their "conventional" counterparts in order to determine whether the novel trait poses altered environmental risks. That is to say the CFIA is interested in whether the PNT presents any new risks not already associated with conventional commercialized varieties in Canada. This comparative approach allows certain risks to be grandfathered in on the basis that their conventional counterparts with similar risk profiles have already been approved. If, for example, existing varieties of a plant pose a threat to biodiversity and the PNT poses no additional risk, the PNT will be considered equivalent to its conventional counterpart and approved regardless of its impacts on biodiversity.

In order to make its assessment the CFIA uses scientific data submitted by the developer of the trait; it does not conduct its own scientific studies on the trait. The CFIA does reserve the right to go back to the developer and ask for clarification or more evidence before it makes its final decision. On top of the environmental assessment a PNT destined for human consumption will trigger a novel food assessment by Health Canada, and a PNT destined for animal feed will trigger a novel feed assessment conducted by the CFIA. These assessments also use the prin-

Figure 6. The regulation of plants with novel traits in Canada

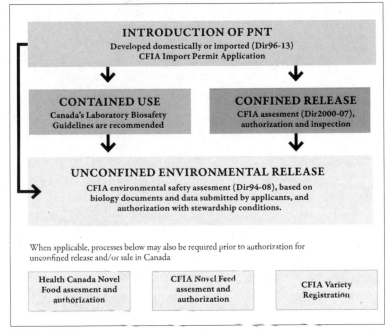

Source: Canadian Food Inspection Agency, http://www.inspection.gc.ca/english/plaveg/bio/pntchae.shtml.

ciple of substantial equivalence to make determinations about feed and food safety. Finally, the variety must be registered by the CFIA upon the recommendation of a crop-specific committee comprised of farmers, plant breeders, quality experts, extension specialists, research centre administrators, and others.

Because the 1993 framework was designed to secure public confidence, a series of public consultations began in the same year and were attended by government, non-government, and civil-society organizations, by academics, and by the private sector. Like future attempts at public consultation, the meetings organized by the newly formed Biotechnology Strategies and Coordination Office were carefully confined to rule out concerns over broader social, political, and ethical dimensions of biotechnology. Instead, only "scientific" approaches to the regulation of biotech products would be heard.

Resistance to the Canadian government's insistence on defining the terms of debate around a "scientific" approach and in treating the products of genetic engineering as substantially equivalent to products of traditional plant breeding developed not only within civil society organizations but also within branches of the government itself. For example, in 1994 the Standing Committee on Agriculture and Agri-Food issued a report on the use of genetically engineered recombinant bovine growth hormone. This report urged decision makers to break out of their "scientific vacuum" and consider the socio-economic and environmental effects of GMOs in order to properly protect the public from "significant risks."[11] Similarly, a report published by the Standing Committee on Environment and Sustainable Development in 1996 highlighted genetic engineering as an inherently ethical issue and thus worthy of special regulatory treatment. This report also recommended that a National Advisory Commission on Biotechnology be established with broad and public representation to consider potential risks, ethical concerns, and the effectiveness of biotech regulations.[12]

Even the National Biotechnology Advisory Committee (NBAC), set up in 1983 to advise the Minister of Science and Technology on the development of biotechnology from the perspective of industry and universities, cautioned the government in its final report in 1998. Citing the lack of public policy tools for "systematically and consistently incorporating socio-ethical considerations" into regulatory and public-policy decisions as well as "no overarching mechanism for holding a non-partisan, national conversation about biotechnology," the report recommended transforming the mandate of the NBAC to an arm's-length advisory committee that would meaningfully incorporate public opinion and socio-ethical issues.[13]

Perhaps the most damning critique of Canadian biotech policy and regulation came from the Expert Panel on The Future of Food Biotechnology convened by the Royal Society of Canada (RSC) at the request of Health Canada, Environment Canada, and the Canadian Food Inspection Agency. The expert panel was made up of fifteen fellows of the RSC with expertise in scientific and policy arenas. The RSC released its report, titled "Elements of Precaution: Recommendations for the Regulation of Food Biotechnology in Canada," in February 2001. As the title indicates, the RSC recommended a precautionary approach to the regulation of GMOs, which challenged the Canadian state's insistence that regulation be based on "sound science." According to Dr. Romeo Quijano, the precautionary approach, favoured by many EU countries, focuses on prevention; reverses the onus of proof of harm by placing this burden on producers; results in the elimination of harmful substances and practices rather than just the management of their associated risks; assesses alternative means to fulfilling human and environmental needs; is scientifically sound; provides unrestricted access to information in an open, democratic, and participatory process; and is based on needs rather than demand.[14] Such an approach stands in stark contrast to the approach of the Canadian government, where risk assessments are based on "sound science." Under the "sound science" paradigm only "objective" scientific evidence is taken into account and, as suggested by Joel Tickner, uncertainty is "frequently considered as a temporary lack of data that can be quantified, modeled, and controlled through additional scientific inquiry."[15] In the "sound science" approach there is no room for an examination of the potential socio-economic implications of GMOs.

In siding with the precautionary approach to the regulation of biotechnology, the RSC made fifty-eight recommendations to change the Canadian regulatory system. Though the government responded to the report with an action plan, political scientists Peter Andrée and Lucy Sharratt of the Polaris Insitute have given the government a failing grade on its implementation of the RSC recommendations.[16] Importantly, Andrée and Sharatt highlight that the government has not heeded the Royal Society's recommendations that a precautionary approach be taken and that the "substantial equivalence" principle be recognized as inherently biased and inadequate. Furthermore, Andrée and Sharatt found no meaningful government action taken to incorporate independent, arm's-length, peer review into regulatory decisions. Instead the CFIA continues to rely on the data provided by the manufacturer of the GMO for its assessment of environmental risk and does not include this data in its publicly available decision documents. In sum, the authors found that many of the Royal Society recommendations have not been implemented because they are in conflict with the government's broader policy commitment to promoting the development and commercialization of biotechnology.

The numerous recommendations and reports released by preeminent societies and advisory bodies such as the RSC and the NBAC, all advocating for increased public participation in Canada's biotech policy and suggesting that socio-economic and ethical concerns be taken into account in regulatory decisions, led the government to pursue an active public relations campaign to restore public confidence in Canada's regulatory apparatus. To that end, the Canadian Biotechnology Advisory Committee (CBAC), created in 1999 under the renewed Canadian Biotechnology Strategy, began public consultations as part of their Regulating GM Foods in Canada Project. As an independent panel of experts, the CBAC was tasked with advising the Biotechnology Ministerial Coordinating Committee on biotechnology policy issues and providing a forum for the Canadian public to engage in a transparent dialogue on biotech issues. In their careful analysis of CBAC's project, geographers Scott Prudham and Angela Morris suggest that CBAC's mandate to directly engage the Canadian public on issues that had so far been beyond the scope of debate, including ethical, social, economic, environmental, and health issues, marked a departure from earlier advisory committees such as the NBAC.[17]

Despite this expanded and participatory mandate, the CBAC was unable to resolve the contradiction it found itself embroiled in—"on the one hand to secure profits and on the other to secure the public."[18] For example, although CBAC was set up as an independent and expert committee, much of the work (including designing and implementing public consultations and drafting key documents) was performed by staff provided by the Canadian Biotechnology Secretariat, a direct affiliate of Industry Canada.[19] Furthermore, the board of CBAC itself was initially weighted toward representation from the business community and the academic and medical sciences research communities. Only one of the twenty board members represented a non-aligned citizen or non-governmental organization. In other words, CBAC's composition reflected the already well-documented problem of the Canadian state, through its various branches, being both a booster of biotechnology and charged with protecting public interests.

Alongside inherent biases in its composition and staffing, Prudham and Morris argue that the CBAC's public consultations amounted to nothing less than "making the market safe for GM foods." Canadian NGOs correctly identified CBAC public consultations as a "participation trap" from the get-go and organized a boycott of the stakeholder workshops, which were planned by a corporate management and accounting consultant, KPMG. Designed to produce legitimacy for Canada's existing approach to the regulation of GMOs, the stakeholder workshops had a narrow scope for discussion and the questioning of the desirability or acceptability of GM foods, which is consistent with a precautionary approach to regulation, was ruled out. After the initial stages of consultation, which included workshops in five Canadian cities and the opportunity to respond to a questionnaire about CBAC's initial consultation document, CBAC released an interim report in the summer of 2001. Canadians were given six months to respond to the recommendations in the report, which was posted on CBAC's website. According to Prudham and Morris, "the committee actually made no changes to its final report on GM food regulation in any way that was explicitly tied to public input."[20] In fact, CBAC did not even make public the comments submitted about the interim report. Although the final report did make recommendations for changes to the regulatory system, they failed to challenge Canada's existing approach to the regulation of GM foods. Instead, the exercise seems to have been more about securing public acceptance of genetic engineering than securing public safety. As sociologist André Magnan concluded about CBAC's GM food study, public concerns about the regulatory system were addressed as problems of poor communication on the part of the regulators and a lack of information among the public.[21] In other words, the state approached the GM food project as an exercise in public relations. This finding supports other research that posits stakeholder dialogues as public relations tactics. Writing on the history of corporate social responsibility, Simon Enoch conceptualizes many stakeholder dialogue processes as "glorified suggestions boxes" meant to subvert new forms of environmental and other activism that use sophisticated communication techniques in the Internet age.[22]

INSIDE THE STATE

Thus far I have presented a picture of the Canadian state's approach to the regulation and promotion of biotechnology as a coherent strategy that has systematically denigrated public concerns and served the interests of the biotech industry through a "science-based" and "product-based" approach aimed at facilitating the rapid commercialization of GMOs. Certainly this picture applies to the key departments and agencies relating to biotechnology, including Agriculture and Agri-Food Canada (AAFC), Industry Canada, and the Canadian Food Inspection Agency. Criticism of the state's approach to biotechnology has, however, surfaced within different sections of these departments and in other arms and institutions of the state, pointing to the need to understand the state as a fragmented and contradictory set of institutions and also as a site of political debate. In other words, in order to maintain a science- and product-based regulatory system, agencies like the CFIA must deal with the sometimes-conflicting mandates of other state institutions and the pressures brought on through parliamentary debate. While thus far the science- and product-based system has remained unscathed, the contingencies associated with the mandates of different parts of the state and the political process itself mean that the future of biotech regulation is not totally fixed.

The two most significant challenges to the product- and science-based regulatory system emanating from within the state and the formal political process centred around (1) the economic impact of adopting GM varieties for farmers and (2) consumer labelling. The first challenge, surrounding economic impact, stems from the historical mandate of Canadian agricultural institutions to promote and protect producer interests. Institutions like the Canada Grain Act, the Canadian Wheat Board, and Agriculture Canada were born out of producer struggle and the state's acknowledgement that in order to secure a settled agricultural economy in the West, farmers would need a reliable minimum standard of livelihood. For example, the Canada Grain Act, established in 1912, emerged out of the Manitoba Grain Act of 1900 and was the product of producers' struggles to protect themselves against the abuses of the large businesses that operated the grain elevators and the handling and marketing systems. True to its origins, the Canada Grain Act's mandate states that it regulates grain handling, maintains standards of quality, and ensures a dependable commodity for domestic and export markets, with the explicit intent of furthering grain producers interests. The Canadian Wheat Board was similarly born of producer agitation and was established as a voluntary government marketing agency in 1935 and soon after, during World War II, gained a monopoly on the export of all Western Canadian wheat for human consumption. And while the CWB act was changed in 1998 to allow "shared governance" between farmers and the federal government, for now even with the recent removal of the single-desk monopoly it is still a quasi-state entity mandated first and foremost to market grain in producers' collective interests. Agriculture and Agri-Food Canada, though not directly tied to producer struggle, has historically understood itself as conducting breeding and research on pests and other agronomic concerns of producers in order to enhance producers' livelihoods. There is, thus, a historical focus within various institutions of the state on developing policies, innovations, and regulations that are aimed directly at protecting producer interests against those of big agricultural business.

In the fight against RR wheat the Canadian Wheat Board was the most vociferous of the (quasi-)state institutions in terms of pushing the issue of economic impact. The CWB's mandate to work in farmers' interests led them to oppose the introduction of RR wheat given that the proper mechanisms were not in place to ensure a positive cost-benefit outcome for farmers. In fact, the CWB advocated that cost-benefit analysis be added as a fourth pillar to the existing food, feed, and environmental safety assessments in the approval process for unconfined release of GM varieties. Most importantly, the CWB joined the coalition to oppose the introduction of RR wheat and was a visible and vocal member of the anti-RR wheat campaign.

AAFC scientists also attempted to intervene in the commercialization of RR wheat. According to documents obtained through Access to Information by a Quebec researcher, in 2001 AAFC scientists brought attention to a clause added to the operation procedures of the Prairie Recommending Committee for Wheat, Rye and Triticale in 1990. The clause stated that "candidates that introduce production or marketing risks for their own or for other wheat classes may be rejected regardless of merit in other traits."[23] AAFC scientists argued that the clause widened the definition of merit used to evaluate a new variety's acceptance beyond agronomic, quality, and disease resistance considerations, and could be used to take socio-economic considerations into account in the committee's recommendations for registration of new varieties. While the role of variety recommending committees is only to recommend new varieties to the CFIA for approval, it would have been largely without precedent for the CFIA to reject a recommendation from the committee. In light of this opening, the CFIA responded swiftly and instructed the committee not to use the clause to influence recommendations since, according to the CFIA, it was not within their mandate to consider socio-economic risks. In 2002 the CFIA changed the operating procedures to remove the clause.

While it is true that the Canadian state is comprised, in part, by institutions that were born out of producer struggle and that undertook the task of supporting viable farm livelihoods, it is also true that neoliberal state policy has eroded, and even eliminated, their capacity to serve producer interests. Elizabeth Moore has traced the changing nature of Canadian agricultural research noting a shift since the mid-1980s toward active promotion of private-sector interests and public-private partnerships in AAFC. She argues that this has caused a fundamental shift away from research that aims to produce public goods (such as research on environmental sustainability and the development of new varieties and practices that benefit producers) in favour of contracts and partnerships with private clients in order to meet explicitly private-sector needs.[24] Rather than developing new publicly funded varieties that would require fewer inputs, such as varieties resistant to drought, disease, and pests, AAFC has increasingly cooperated with private companies that patent varieties and lock farmers into expensive and proprietary agronomic input packages. Monsanto's Roundup-resistant crops are perfect examples, as they require the application of the Roundup herbicide and force farmers into the marketplace each year to buy new seed.

The Canadian Wheat Board has also come under attack by neoliberal state policy. Most dramatically, the Conservative Party majority government elected in 2011 made eliminating single-desk marketing among its top legislative priorities. In October 2011, just weeks into the first session of parliament in its majority mandate, the government introduced Bill C-18, *Marketing Freedom for Grain Farmers Act*. This legislation, which ended the CWB's monopoly, took effect on 1 August 2012. Members of the farming community have read this bill as a heavy-handed, ideological, and even illegal attack on collective marketing structures. For example, the government limited parliamentary discussion on the bill immediately after introducing it and insisted that the bill speed through parliament by mid-December of 2011.[25] In response, the CWB itself filed a federal lawsuit on 26 October 2011 that challenged the legality of the legislation. The CWB argued that the Canadian Wheat Board Act requires that changes to the single-desk structure be first approved by a vote of affected producers and that the CWB board of directors be consulted; neither of which had been fulfilled.[26] A federal court sided with the CWB in a decision in December 2011, but the government went ahead with the legislation anyway and launched an appeal to the federal ruling. In June 2012 the Federal Court of Appeal overturned the Federal Court's December ruling. At the time of writing eight former farmer-elected Directors of the CWB were seeking to appear before the Supreme Court of Canada.

The issue of market impact has not only sprung up in the various institutions of the state with mandates to pursue producer interests but has also been the focus of the formal political process including political debate in the House of Commons. For example, the *Western Producer* reported in late 2003 that then-Minister of Agriculture Lyle Vanclief was "working with industry to create a 'new step' in the process of bringing new varieties to market, a step that recognizes such issues as consumer attitudes and potential market impact."[27] Responding to Liberal backbencher Charles Caccia (a vocal critic of GMOs) in the House of Commons, Vanclief assured him that he was working on adopting a mandatory step in the approval process for GMOs over and above the existing safety regulations. Such an additional step never saw the light of day, since Monsanto withdrew its application in May 2004, just a month before the federal election. Furthermore, Minister Vanclief had already announced he was retiring from politics at the end of 2003, only a month after disclosing his intentions to institutionalize market impact as part of the regulatory process. Vanclief's intentions, though ultimately abandoned, do point to the more contingent nature of legislative institutions of the state, whose members are elected and, thus, subject to public campaigns and the pressures of civil society.

Well after Monsanto pulled its application for commercial release of RR wheat, Bill C-474, *An Act Respecting the Seeds Regulation,* was introduced in the Canadian Parliament in March of 2010. The private member's bill was meant to deal a blow to the impending registration of GM alfalfa (Monsanto's RR alfalfa was approved for environmental release in 2005, but Monsanto had not yet applied for variety registration of the product) and to prevent the introduction of other GM crops including the possibility of Monsanto reintroducing RR wheat. Introduced by the New Democratic Party's agriculture critic Alex Atamanenko, the bill would have required an analysis of potential harm to export markets before the commercialization of any new genetically engineered seed. The bill made it through several readings in the House and was referred to the Standing Committee on Agriculture and Agri-Food for further hearings. According to the Canadian Biotechnology Action Network, such a long and engaged debate on genetic engineering is without precedent in the House of Commons.[28] Furthermore, the debate proceeded despite attempts by the biotech industry to thwart its airing in the House. In February 2011 bill C-474 was finally defeated during its third reading.

The second most significant challenge to the product- and science-based regulatory system emanating from within the formal political process centred on the issue of consumer labelling. From 1999–2003 several private member's bills from both government and opposition benches were introduced in Parliament, calling for mandatory labelling of foods containing GMOs. Charles Caccia, a backbencher in the governing Liberal Party and chair of the Standing Committee on the Environment and Sustainable Development, was particularly active in this regard. In every parliamentary session between 2000 and 2003 he introduced legislation requiring mandatory labelling.[29] His 2001 private member's bill came close to making legislative change but was voted down 126 to ninety-one and thereby prevented from going to a second reading. Key Liberal MPs who supported consumer labelling, such as Health Minister Alan Rock, were absent during this vote. A similar bill, introduced by Caccia in 2002, gained considerable support in the House but narrowly missed a new rule that would have sent all private member's bills worthy of House debate to a vote. Instead, a committee of MPs decided the bill would be debated, but not voted on, in December, just a month before the new voting practice was to be implemented.[30] As reported by Barry Wilson in the *Western Producer*, MPs in both government and opposition considered a vote on this bill too close to call.[31]

Despite Caccia's persistent attempts to bring about mandatory consumer labelling through the formal political legislative process, the issue was ultimately decided through a committee of public- and private-sector representatives. Back in 1999 a committee established under the Canadian General Standards Board had been tasked with developing guidelines for voluntary labelling of GM foods. The work of this committee was mired in controversy, with several environmental and other

NGOs refusing to take part because mandatory labelling had been precluded from the beginning. Despite widespread consumer support for mandatory labelling, the committee finally agreed to voluntary standards in 2004, and the issue of mandatory labelling seems to have disappeared from the formal political process.

CONCLUSION

Since the introduction of biotechnology, the Canadian government has made a consistent set of choices in the development of its biotechnology policies. From the beginning the Canadian state has actively promoted the development of the biotech industry as part of a strategy of international competitiveness and innovation. This early promotion has led the government to pursue a regulatory approach that aims to foster a secure environment for biotech firms by shielding the regulatory process from public concerns over matters such as social, ethical, and economic risks. The main mechanisms through which the state maintains this shielded regulatory process is through its "science-" and "product-based" approach wherein social, ethical, and economic risks are defined as unscientific and where products of biotechnology are considered substantially equivalent to their conventional counterparts. When the legitimacy of the science- and product-based approach has been under threat, the government has attempted to assuage the concerns of citizens, producers, and consumers through exercises in public relations—consultative processes that, in practice, have acted as participation traps.

According to political scientists Sarah Hartley and Grace Skogstad the Canadian state has been able to pursue its product- and science-based approach without fatal challenge by minimizing the role of the institutions of representative democracy in favour of a strategy of functional democracy.[32] So while issues surrounding the economic impact of RR wheat and consumer labelling have been subjects of parliamentary debate and parliamentary committee studies, elected politicians have had minimal involvement in Canada's policy and regulatory framework for agricultural biotechnology. In fact, all formal political debate about GMOs has come late in the game because of the government's early decision to use the existing Seeds Act to regulate the products of biotechnology. The little parliamentary debate that has occurred has thus far been unsuccessful in securing regulatory changes.

Rather than subjecting biotech policy and regulation to the institutions of representative democracy, the Canadian state has relied on advisory bodies and stakeholder consultations to develop biotech policy. Under this "functional democratic" approach, the state has maintained a closer hold on the policy making process by identifying the relevant stakeholders and, thus, constraining the process to a discourse of competitiveness and innovation wherein only "scientific" approaches are considered. When such bodies have recommended that the government diverge from its promotional practice and its science- and product-based approach (most forcefully in the Royal Society of Canada's 2001 report advocating for a precautionary approach to the regulation of GMOs), the state has not acted on these recommendations.

It is true that the state can be a fragmented and contradictory set of institutions. The fate of GM crops is, therefore, not forever determined. Owing to their historical ties to producer struggle, some institutions of the state are mandated to protect producer interests against exploitation from big business. However, such institutions have come under systematic attack by neoliberal policy and governments intent on destroying collective marketing arrangements and opening up public institutions to private interests. If a serious divergence from the science- and product-based approach is ever to come about in Canada, producer-centred government institutions will be required to play a leading role. Otherwise, producers will have all the more trouble tackling the promotional stance of the state without the Canadian Wheat Board, the Canadian Grain Commission, and public scientists freed from private partnership and private funding.

The Difference
BETWEEN BREAD AND OIL
People-Plant Relationships in Historical Context

CHAPTER THREE

Before the controversy over GM wheat heated up in Canada, prairie producers had been relatively quiet on the politics of biotechnology. Meanwhile, their counterparts in the milk producing regions of Quebec and Ontario were engaging, alongside consumer and health advocates, in a fairly public campaign to oppose the regulatory approval of Monsanto's recombinant bovine growth hormone, which Health Canada ultimately rejected in early 1999. In the late 1990s urban, health, and consumer movements had also been involved in a campaign for mandatory labelling of GM foods, which, despite highly politicized debates in the House of Commons, was ultimately unsuccessful. It was not as if prairie farmers had been left out of the biotechnology "revolution"; an increasing number of them had been growing genetically engineered herbicide-tolerant varieties of canola since 1995, and a number of private companies and public research programs were aggressively pursuing genetic modifications to a whole spectrum of prairie field crops. Why then was it not until the threat of RR wheat hitting the market that prairie farmers publicly voiced their opposition en masse?

While the answer to this question is complex, in this chapter I contend that the contemporary and historical political ecologies of canola and wheat provide insights into the diverging politics that surround their genetic modification.

THE COPRODUCTION OF HUMANS AND NONHUMANS

Long before the advent of genetic engineering farmers and plant breeders had already thoroughly transformed wheat and canola. In fact, Lawrence Busch, William B. Lacy, and Jeffrey Burkhardt claim that "it would not be an exaggeration to say that the present-day wheat plant has been 'socialized.'"[1] That is to say, humans have not only transformed the biological properties of wheat and canola but also subjected these plants to a whole host of social relations characteristic of the societies of which they are part. Social relations such as private property regimes, cooperative marketing structures, and public and private breeding programs have left their imprints on the characters of the wheat and canola industries, albeit in dissimilar ways. However, any narrow focus on the ways in which humans have transformed the natural and social relations of wheat and canola ignores the ways in which wheat and canola have participated in the process of transformation—how humans have also been shaped through their relationships with these plants. In order to avoid an over-socialized account of the histories of wheat and canola, attention also needs to be given to the role of canola and wheat as material entities that have shaped and constrained the process of the humans' and plants' mutual "socializaton."

Throughout her many works, Donna Haraway has been centrally concerned with the role of non-humans in "social" life and the ways that humans understand and represent the non-human others that co-inhabit the world. In this spirit, much of Haraway's work is devoted to developing representations (or *figurations,* as she calls them) of hybridity with the goal of deconstructing binaries such as object/subject, organism/machine, and nature/culture. For example, she uses a coyote or trickster figure to represent nature as active subject(s) capable of subverting human intention.[2] In this conception, nature is not the sole production of humans, nor is it pre-existent and unchangeable. Rather, it is a co-production in which non-humans play an important part; they are not simply objects to be manipulated by humans. The reconceptualization of nature as active rather than passive through the coyote figure explicitly disrupts what Haraway calls "nature/culture ontologies."[3]

Haraway is not unique in her insistence on attending to the role of nature in shaping the world. In fact, social scientists have increasingly attempted to recognize that the "hard" material form and properties (also called "the materiality") of nature constrain and shape the ways in which nature is engaged and worked up by humans.[4] More controversial has been the theorization of nature as active agent, inspired largely by scholars of science and technology studies.[5] In his study of techno-politics in twentieth-century Egypt, Timothy Mitchell foregrounds the ways in which human intentionality and agency depend on attaching themselves to or redirecting non-human energies and logics. For Mitchell, human agency is partial and incomplete, the result of interactions with non-human others including mosquitoes and technologies. In examining the complicated relationships between war, famine, and malaria that led to the disaster of 1942–44, Mitchell shows how the invasion of a new strain of malaria was the result of unforeseen combinations of human intentionality and non-human energies, including failing agricultural crops, changed hydraulic landscapes, weakened immune systems, and new routes for mosquitoes carrying the *Plasmodium falciparum* parasite.[6]

Drawing on the work of science and technology studies scholars, geographers using actor-network theory have similarly posited non-humans as potential agents. For them agency is relational; it is not located in individual subjects, but arises out of the interactions or associations of a plurality of humans and non-humans.[7] Importantly, this conception of agency does not require consciousness or intentionality. Rather, agency and power are here understood as radically decentred since they are no longer located in individuals but arise out of human and non-human interactions. Therefore, it becomes just as important to attend to microbes, technologies, and plants and animals as to institutions and capital. In this respect a researcher cannot assume the priority of society over non-human relations.

Haraway's approach nicely balances attention to the agency of non-humans and a focus on structural relations of power, including the systematic logics through which capitalism operates. Importantly, Haraway is also interested in the ways that ascribed meanings and symbolism affect how social relations play out on the ground. According to Haraway "there is no gap between materiality and semiosis; the meaning-making processes and the materiality of the world are dynamic, historical, contingent, [and] specific."[8] On the one hand Haraway highlights that objects/actors/subjects have a "real" physicality or materiality, while on the other hand she recognizes that objects are in part constructed through social interaction and that symbolism and representation (or semiosis) do important work in the operation of power.

In this chapter, wheat and canola are understood as both material and semiotic (i.e., the product of meaning-making and symbolism), and as playing some sort of active role in constraining and enabling human action. Haraway's more recent figuration of companion species is also useful in thinking about the material and semiotic histories and co-evolution of wheat, canola, and people. In *When Species Meet*, Haraway digs into the long history through which humans and dogs came to live with one another. She shows how this process involved a co-evolution in which dogs were domesticated, but humans were also thoroughly changed. The book is concerned with the whole history of human/dog co-evolution, from the domestication of dogs from wolves to contemporary U.S. dog culture. Here Haraway is keen to show how science, including veterinary science and genetics, has been instrumental in shaping dog-human relationships and developing and defining breeds, both semiotically and materially.[9] Although Haraway uses the dog as her example, companion species also include "such organic beings as rice, bees, tulips, and intestinal flora, all of whom make life for humans what it is—and vice versa."[10] It is thus perfectly reasonable that wheat and canola might be considered companion species of Canadian farmers and eaters. Through the examples that follow it is quite clear that the histories, identities, and biologies of Canadians, of eaters, and of prairie farmers are tied up in the history of wheat and canola, and vice versa. As I will show, though wheat and canola can both be successfully understood as companion species, they are radically different companions embodying dissimilar materialities and semiotics.

. . . . W H E A T . . . .

"Wheat is 14 percent protein and 86 percent politics."
— *Cited in Julian Thomas, "A Technical Critique of the Western Canada Quality Assurance (QA) System in Wheat"*

ORGANIZING THE FRONTIER

The colonial settlement of the prairie provinces of Canada, beginning in the late 1800s, was predicated upon the establishment and growth of a white settler wheat economy.[11] During this time, prairie ecology and society was almost completely (re)arranged around the household production of wheat for export. Meanwhile, Indigenous peoples were forced onto reservations with no access to homesteads. Their attempts at farming were actively constrained by the colonial administration through the limiting of crops and machinery so as to thwart competition with the settler wheat economy.[12] Millions of migrants were attracted to the provinces of Manitoba, Saskatchewan, and Alberta by the promise of ten-dollar homesteads[13] and the prospect of profitable participation in the emerging global wheat economy.[14] From a combined total of 419,512 in 1901, the population of the three prairie provinces grew to 1,328,121 in 1911, broke the 2-million mark by 1926, and continued to grow with less fervour after that.[15] The vast majority of these new migrants did indeed settle on farms. In 1931, for example, agriculture accounted for more than 60 percent of direct employment in Saskatchewan, 50 percent in Alberta, and 34 percent in Manitoba, while the national average was only 28.73 percent.[16] Furthermore, wheat accounted for the majority of field crop production in the early 1900s, ranging from 55 to 68 percent of seeded acreage (Table 4). During this period wheat served as the primary cash crop, while other grains were grown to feed farm animals. Despite a more diversified contemporary farm economy, wheat still accounts for a considerable portion of production (over one-third of all seeded area in 2007, see Table 4) and is grown on roughly half of all prairie farms (see Table 5).

The development of the prairie wheat economy was essential not only to the settlement of the prairies but also to the 1879 National Policy that sought to develop an integrated national economy in the newly established Confederation. As Vernon Fowke explains in his highly regarded analysis *The National Policy and the Wheat Economy*, "the prairie provinces constituted the geographic locus of the Canadian investment frontier in the first three decades of the twentieth century. The dynamic influence of the frontier permeated and vitalized the Canadian economy and extended far beyond."[17] Specifically, the settlement of the prairies and the establishment of the wheat economy necessitated huge amounts of capital that the National Policy, through tariff protection for Canadian manufacturing, ensured would be produced in and supplied by central Canada. The tariffs served to build and protect a spatially

Table 4. Seeded area (acres) in the prairie provinces

YEAR	TOTAL PRINCIPAL FIELD CROPS SEEDED (ACRES)	TOTAL WHEAT SEEDED (ACRES)	WHEAT AS PERCENT OF AREA SEEDED	TOTAL CANOLA SEEDED (ACRES)	CANOLA AS PERCENT OF AREA SEEDED
1910	13,001,000	7,867,400	60.5	0	0
1920	30,688,800	16,841,200	54.9	0	0
1930	39,001,600	23,960,000	61.4	0	0
1940	40,583,300	27,750,000	68.4	0	0
1950	41,823,400	26,382,000	63.1	400	0.001
1960	41,228,200	23,900,000	58	763,000	1.9
1970	36,870,500	12,000,000	32.5	4,050,000	11
1980	49,317,000	26,900,000	54.5	5,000,000	10.1
1990	57,480,000	33,780,000	58.8	6,120,000	10.6
2000	59,467,000	26,352,000	44.3	12,050,000	20.3
2007	55,915,000	20,820,000	37.2	14,450,000	25.8

Source: Statistics Canada, *Table 001-0017—Estimated areas, yield, production, average farm price and total farm value of principal field crops, in imperial units, annual,* CANSIM (database), using E-STAT (distributor).

Table 5. Percentage of farms growing wheat in the prairie provinces

YEAR	MANITOBA TOTAL # FARMS	MANITOBA PERCENT WHEAT	SASKATCHEWAN TOTAL # FARMS	SASKATCHEWAN PERCENT WHEAT	ALBERTA TOTAL # FARMS	ALBERTA PERCENT WHEAT
1991	25,706	63%	60,840	84%	57,245	41%
1996	24,383	51%	56,995	75%	59,007	34%
2001	21,071	45%	50,598	66%	53,652	29%
2006	19,054	38%	44,329	55%	49,431	24%

Source: Statistics Canada, 1991, 1996, 2001, and 2006 Census of Agriculture, data by province, Census Agricultural Region (CAR) and Census Division (CD) (database), using E-STAT (distributor).

concentrated, highly subsidized manufacturing economy in central Canada. In this policy context, the settling of the prairies provided domestic markets for agricultural machinery, tools, hardware, home furnishings, and textiles, since each prairie household production unit needed several pieces of farm equipment, and materials for building a house, barn, granary, etc. Furthermore, new industries in grain handling, flour and baking, and meat packing required more railway routes, and the development of trade centres, warehouses, elevators, and much more.

Populist political organizing began on the Canadian prairies as it was settled. Farmers contested the ways in which the national economy was being built and the position of the yeoman farmer within that system. Prairie farmers located their exploitation in two main areas. First, they were particularly upset about the tariffs of the National Policy, which forced them to buy machinery, farm inputs, and household supplies in protected markets but to sell their grain in open international markets. Second, they blamed the capitalist economic system for monopolies in the railroad, elevator, grain, and banking industries and for international cartels in the fertilizer/agro-chemical, grain-handling, flour-mill, and farm-machinery industries. For example, railways and elevators colluded so that producers were forced to deliver to their local elevator, which assigned dockage, grades, weights, and prices that cheated the farmer.[18] In short, farmers felt that they were barely able to earn enough from the sale of their products to buy the required inputs for the upcoming year and to meet the minimum requirements for familial consumption.[19] Any potential surplus (over and above what would be needed to buy the factors of production and feed the family) was being completely siphoned off by input and credit suppliers upstream or grain merchants, elevator companies, and the railway downstream.

Despite their relative atomization on distant homesteads, farmers began to organize collectively against the extreme extraction of surplus (i.e., the ability of upstream and downstream commercial actors to capture all the potential farm profit) from farming. Immigrants from Britain and the United States brought with them experience in populist and cooperative organizing, and they put this to work on the Canadian prairies. Political organizing was decidedly anti–big business, intent on keeping private ownership of the means of farm production—including land and farm machinery.[20] Early unrest sparked the first royal commissions on grain handling and led to the establishment of the Manitoba Grain Act in 1900. This regulatory framework established a legal basis for farmers to bypass the unfair treatment they received from elevators and railways; it required railroads to supply cars directly to producers and elevators to establish platforms from which farmers themselves could load their grain. In 1901 producers organized the Territorial Grain Growers' Association (renamed the Saskatchewan Grain Grower's Association in 1905 with the establishment of that province), which became an active and powerful voice for farmers. In fact, according to the association's first president, W.R. Motherwell, farmers were quite close in the early 1900s to becoming a violent force, one that provincial or federal politicians would not be able to ignore.[21] The Grain Grower's Association was adamant that the Manitoba Grain Act be enforced and, in a highly publicized trial, it fought a Sintaluta, Saskatchewan station agent for not granting farmers' railcar orders in 1902. This farm organizing was happening in the context of broader politicization around work in Canada. Especially in Western Canada, immigrants from Central and Eastern Europe were drawing on social democratic and socialist traditions from their home countries to fight the exceptional exploitation they were experiencing from their employers and from their Anglo co-workers in the early 1900s.[22]

Much of the early farm organizing on the prairies took the form of building producer cooperatives in the grain handling industry, including cooperative elevator companies in Saskatchewan and Alberta and a prairie-wide marketing agency established in 1906 called the Grain

Growers' Grain Company (GGGC). For a short time the GGGC had an office in Winnipeg and traded on the Winnipeg Grain Exchange, until it was displaced for breaking the rules of the exchange by returning patronage refunds to its members. Despite this setback, the GGGC was considered very successful and popular with farmers; as many as 9,000 were marketing their grain through the company in 1910.[23] Despite incomplete coverage in Western Canada, the GGGC marketed over 16 billion bushels of wheat in the 1909–1910 crop year, which accounted for 15 percent of the total wheat crop in that year.[24] Meanwhile, the Saskatchewan Grain Grower's Association was demanding the socialization of elevators in Saskatchewan. After a commission examined the issue, the Saskatchewan government ultimately decided against a government-owned elevator system, but came out in favour of cooperative ownership with the backing of low-interest capital from the province.[25] This served as the model for the establishment of the Alberta Farmers' Cooperative Elevator Company in 1913.[26] As Paul Sharp highlights in *The Agrarian Revolt in Western Canada,* "though the grain growers' societies considered themselves 'non-partisan, non-political, non-trading,' they were class-conscious organizations and concerned themselves with marketing problems from the beginning."[27]

The struggle to organize and maintain pools in wheat was another great success for prairie producers. They had had a taste of the power and effectiveness of pooling in wheat when a short-term government board was given monopoly power during the First World War. This provisional board was created in order to deal with increased demand from Europe and the uncertainties inherent to the Winnipeg Grain Exchange as prices in wheat rose and the futures market took off. After the war, farmers called for the continuation of this board, but decided to take matters into their own hands when the federal government returned wheat to the open market. In the early 1920s, through large-scale and labour-intensive drives that involved going door-to-door to convince farmers to sign contracts promising that they would pool their wheat, the Manitoba, Saskatchewan, and Alberta wheat pools were established. In Saskatchewan it took two attempts to enrol 50 percent of the total wheat acreage, below which contracts were invalid, while Manitoba and Alberta went ahead with less than 50 percent of the total wheat acreage in 1923 and 1924 respectively.[28] In 1924 a central selling agency was established by the three pools and soon thereafter the pools began building and buying elevators and port terminals through which to conduct their business. Garry Fairbairn's biography of the establishment of the Saskatchewan Wheat Pool (SWP) claims that: "At its birth SWP was a radical, almost revolutionary, attempt to replace the existing order. Pool founders did not want to improve the Grain Exchange; they wanted to eradicate it. They did not want to become respectful, influential members of the traditional free-enterprise system; they wanted to create a co-operative alternative to that system. Regardless of whether cooperation was considered as the highest form of free enterprise or as some moderate form of socialism, the ultimate goal was to see co-operation spread through society."[29]

When international grain prices collapsed in 1929 and the Great Depression began, the Pool's Central Selling Agency found itself in huge financial difficulty, owing substantial sums to banks and having already given out initial payments for the 1929 crop that international grain prices could not support. Farmers' demands to bring back a national, state-operated central selling agency and pool for wheat (which had been a temporary measure during the First World War) were finally heeded. In 1935, by an act of Parliament, the Canadian Wheat Board was re-established as a voluntary marketing agency for wheat. The provincial pools continued to operate their cooperative grain handling systems, allowing farmers to bypass the private grain companies. Cooperatives were also used to fight against banking monopolies; credit unions and cooperative insurance began to be organized in the late 1930s.

The narrative I have provided here certainly glosses over many of the nuances of a dynamic and complex history. Indeed, dozens of manuscripts exist on each of the significant events and processes described above. Furthermore, important political distinctions were beginning to develop between the different provinces' approaches to cooperation and struggle. Interested readers will have to seek these out elsewhere. The point here is to highlight that wheat occupies a prominent place in Western Canada's political history and the stories Western Canadians tell about their pasts. First, a major thrust of the political economy of Canada was oriented toward opening up the West to settlement and participation in the wheat economy, the legacy of which continues to shape current politics and economy in the prairie provinces. Second, wheat was the crop around which producers organized cooperatively in order to curb the extreme extraction of surplus from farming by large corporate players. Wheat became both a powerful material and symbolic expression of a strong independent and mobilized class of farmers.

THE INDUSTRY: MONOPOLY, BREEDING, AND PRIVATE DIVESTMENT

Western Canadian agriculture is structured around the specific political, institutional, ecological and cultural characteristics, and histories of particular crops. The wheat industry, for example, has a unique historical and institutional character that presents certain opportunities and constraints for the future. Perhaps most prominently, the political organizing outlined in the previous section resulted in the establishment of a single-desk government marketing agency for wheat. The Canadian Wheat Board is premised on maximizing returns to producers and has played a powerful role in wheat commodity chains. The wheat industry is also characterized by the historical investments made by the Canadian state in breeding and research, which, as my research participants noted, are currently being eroded. As a low-value commodity with a history of public investment and relatively unrestricted access to seed and seed saving, wheat, unlike canola, has received less interest and investment from private capital. It is to these particularities of the industry that I now turn.

The pooling of wheat, which has taken place on the prairies since the early 1920s (first through the provincial wheat pools and then through the Canadian Wheat Board), has allowed producers to have some influence in the wheat industry and to manage the marketing of wheat for their collective economic interests. Originally established in 1935 as a voluntary, government agency for the marketing of wheat, the Wheat Board gained increasing clout and influence as it was called on to smooth over tumultuous economic and political crises. During World War II, for example, the CWB mandate was extended to the marketing of all Canadian grains. It was also during World War II that the government granted the CWB monopoly power, requiring producers to deliver to the board. Rising prices due to the war in Europe were tempting producers to sell to private grain merchants who could offer spot prices that exceeded the initial payments of the CWB.[30] Thus, if wheat was going to continue to be marketed for the collective good, even during times of relative prosperity, delivery to the CWB would have to be compulsory. In 1949 the CWB's mandate was narrowed from the marketing of all grains to a monopoly of wheat, oats, and barley.

Over the last fifty years the scope of the CWB was further narrowed, and on 1 August 2012, its single desk marketing monopoly was removed through legislation.[31] Importantly, in 1998, the CWB Act was changed to allow "shared governance" between farmers, who elected ten of the fifteen members of the board of directors, and the federal government, which appointed the other five, including the president and CEO. Until August 2012, the CWB played a very influential role in the Canadian wheat industry, conducting business for 70,000 Western Canadian farmers.[32] The fate of the CWB in a post-single-desk era is as yet unknown. For example, it will likely lose much of its considerable influence on breeding agendas at Agriculture and Agri-Food Canada, in developing knowledge and technologies that are useful for advancing quality characteristics of interest to wheat processors through the Canadian International Grains Institute, in consulting regularly with the Canadian Food Inspection Agency about any changes to regulations and variety registration, and in negotiating constantly with players in the grain handling and transportation industries. Before August 2012 it maintained close ties to government and was consulted in a meaningful manner about policy and other changes to the wheat industry. In this way, the CWB ensured that farmers' interests were represented throughout wheat commodity chains and that the organization retained a central role in the industry.

The wheat industry is also characterized by the Canadian state's involvement and investment in breeding and research. A high rate of crop failure (due to drought, frost, disease, etc.) and of abandonment of homesteads in the early years of settlement forced the federal government to ameliorate the conditions of settlers through initiatives such as experimental farms meant to develop new plant and animal crops and breed varieties suited to local climatic conditions. At the first of such farms, established at Indian Head, Saskatchewan in 1887,[33] and at other stations, researchers focussed their efforts on wheat and developed and tested new methods to control pests, disease, and weeds. They also began experimenting with breeding to find an earlier maturing variety. By 1907 the breeders at Indian Head had introduced the famous Marquis variety. This early maturing, high-protein wheat was much better suited to the Canadian prairies and allowed for the spatial expansion of wheat production.[34]

In her 2002 analysis of changing federal agricultural research agendas and practice in Canada, Elizabeth Moore emphasizes that, until recently, the federal Canadian state has been the dominant actor in agricultural research, with the goal of increasing production capacity among farmers and ensuring a profitable agri-food sector.[35] Early research at regional stations addressed the needs of and problems experienced by farmers, while later (by the late 1960s) federal scientists focussed on efficiency in production. According to many of my research participants, this history of public investment and the prominence of non-capitalist organizations like the CWB in the wheat industry has resulted in less private investment in wheat than in other crops. For example, a representative from the Western Barley Growers Association, an organization that is publicly opposed to the CWB's monopoly, explained that the wheat industry is in decline. According to him, wheat used to account for 80 percent of the cash receipts of prairie farmers, but that figure has been reduced to 20 percent today. In an interview with me he blamed a strict regulatory system, including the CWB, for stifling innovation in wheat and therefore reducing Canadian farmers' share of the growing global wheat market.

Moore identifies significant changes to the federal approach to and investment in agricultural research post-1980. Most significantly, the Research Branch of Agriculture and Agri-Food Canada (AAFC) has begun to orient its research to the needs of the agri-food industry, and has suffered from budget cutbacks. Moreover, scientists from AAFC suggested that private firms have given much less attention to wheat than they have to other crops, especially at the level of breeding. Comparing canola to wheat breeding, a scientist at AAFC told me that private industry simply cannot turn a profit in wheat breeding so they have left it largely to public sector institutions. According to this scientist there are roughly fifty new cultivars per year in canola, which provides ample opportunity for private profits from plant breeder's rights, while the public sector produces only five to eight new cultivars per year in wheat. The number of wheat breeders in Canada has declined to seven—all of whom are primarily working within public institutions such as AAFC. In the context of both public and private divestment, farmers have begun to fund producer-centred research programs at AAFC through mechanisms such as voluntary check-offs (sums of money deducted from cash payments for crops that are directed toward research). The Western Grain Research Foundation, established in 1993, is an example of a non-profit organization that funds research and breeding in wheat and barley through producer check-offs.

Low commodity prices in wheat, compared to newer higher-value crops, have also resulted in less willingness and ability among prairie farmers to pay for corporate wheat seed. Furthermore, farmers have had a long history of fairly easy and relatively unrestricted access to wheat seed through public breeding and distribution programs and seed saving. According to a scientist at AAFC only about 15 percent of all cereal grains in Western Canada are sold as certified seed. This means that farmers are escaping royalty payments on roughly 85 percent of the seed that they plant. As an interviewee from Agriculture and Agri-Food Canada explained to me, farmers are thus extra sensitive to royalty rates in wheat; if rates get too high farmers will opt to reproduce seed on their farms. Indeed, seed saving in wheat was practised by all of the farmers that I interviewed, even if they did not save seed from other crops. Thus, despite the introduction of plant breeders' rights in Canada in 1990 (which legally enshrine a plant breeder's exclusive control over new varieties for up to eighteen years) private firms fear that a culture of seed saving and low market prices for wheat will not leave enough opportunity for private accumulation.

Significant changes to the structure and character of the wheat industry are likely over the next few years. While public divestment of the nature described by Moore is likely to continue, new avenues for private profit-making may be in store. For example, as biofuels gain increasing prominence both domestically and internationally, new opportunities in processing wheat for biofuel, changes to breeding agendas, and variety registration will open up new areas to private investment. Furthermore, the end of the CWB's single desk opens up the wheat trade to significant opportunities for private profit-making. Mergers and acquisitions in the grain handling business (for example, the recent sale of Viterra to Glencore International) are already afoot with more to be expected in the near future. The legacy of public investment and the involvement of non-capitalist organizations in the wheat industry are clearly under assault; their decline will continue to structure the speed and manner in which private interests engage with wheat. It is sure that farmers and the broader public will continue to struggle for the public and collective character of the wheat economy.

IN THE FIELDS
Wheat has certain biological and agronomic characteristics that have made it a crop that is particularly attractive to prairie farmers. While seemingly innate and fixed, these properties are co-produced by the farmers, scientists, and organisms that are involved in the breeding and cultivation of wheat. In other words, breeders have selected the traits they have found most beneficial, literally shaping the biology of wheat; farmers have adapted wheat to particular cropping systems, and their agronomic practices to the needs of wheat plants. Specifically, public investment in wheat breeding since the late eighteen hundreds has produced varieties that are particularly well-adapted to the short and dry growing season of the Canadian plains. Furthermore, the farmers whom I interviewed as part of this research report using wheat in their rotations to break cycles of disease. In addition, a variety of fairly inexpensive pesticides have been introduced to effectively control weeds in wheat crops. For these reasons wheat continues to be a very important part of rotations and prairie farm economies even though farm prices for wheat are low compared to other field crops like canola, lentils, and peas. While wheat no longer dominates production like it did in the early 1900s, it remains a widely planted crop among farmers.

It was not a coincidence that Canada's hinterland economy was developed almost singularly around wheat. Importantly, wheat was in demand the world over; but as significantly, "intense specialization in wheat relative to other grains on the semi-arid plains [was] a result of the great drought-resisting capacity of wheat and the extremely high quality of the wheat produced."[36] While wheat yields were, and still are, higher in moister and warmer parts of the country (for example in Southern Ontario) the prairies continue to dominate in the production of wheat. This is because wheat is one of the few crops that is easily grown on the prairies at very high quality standards, and because the more fertile regions of Canada are able to grow a greater diversity of more lucrative crops. A long history of public plant breeding programs in wheat has made the crop even more adapted to the prairie environment and has allowed for the extension of the wheat growing area.[37] Since weather and climatic conditions continue to be a great source of risk for prairie producers, many grow some wheat every year in hopes that it will survive even if other crops fail.

Representatives of farm organizations who participated in this research characterized wheat as a valuable crop, despite the fact that it commands a relatively low price. Wheat is valuable to farmers because of what it does in the fields and the possibilities it enables in terms of crop rotations. For example, this producer from the Keystone Agricultural Producers, whose experience was echoed by many others whom I interviewed, places more importance on wheat's agronomic value than he does on its economic value: "Wheat is the one crop that we grow, we grow it hoping to make money; we usually break even in years like this. It breaks the disease cycle; you can grow wheat after wheat and not have too many disease problems. But if you grow canola after canola you could very easily and likely lose the crop. So we need to grow some wheat and other crops to break the cycle. Cereal crops break the cycle of weeds."

While a staple in nearly every farmer's rotations, wheat is especially important to organic farmers. This representative of the Saskatchewan Organic Directorate explains that wheat is "a key crop to us. It's easy to grow, relatively easy to grow. It's weed resistant, it's usually easy to market, there's a whole host of reasons why it's a good crop. It's something that every organic farmer pretty well grows."

Agronomy and biology are also fundamental to the work of plant breeders and companies developing agro-chemicals and agricultural biotechnologies. The possible biological selections, manipulations, and chemistries are limited by the actual agronomic practices of farmers across space. This tension between what is biologically or chemically possible and what is agronomically practical has taken on a more overtly political form in the debate around genetic modification. For example, what is considered an acceptable modification in one crop may not be suitable in another. One scientist working at a public university explained that the insertion of "risky" traits into organisms that are highly outcrossing (i.e., prone to mating with dissimilar genetic strains) raises uncertainties that may be unacceptable to farmers and the general public. For example, if a herbicide-tolerant canola were to cross with a wild relative and transfer its resistance, farmers would have to contend with a very difficult weed. Since wheat has a much lower incidence of outcrossing than plants like corn and canola,[38] the main concern regarding the interplay of biology and agronomy, in this case, has centred on the utility of traits such as herbicide resistance in a crop that is already effectively managed by farmers. Even plant breeders who supported the introduction of herbicide resistance in canola publicly voiced their opposition to its insertion in wheat for precisely this reason, as this interview participant from Agriculture and Agri-Food Canada demonstrates: "If weeds were a real problem in cereals and we didn't have the herbicides that would effectively and cheaply control this then I would have said okay, Roundup Ready [herbicide-resistant] wheat is a good idea. But the fact of the matter is that we have some very good chemicals for cereals that are relatively cheap and relatively benign and so it's not essential that we have Roundup in it."

The examples above serve to illustrate how the biology of prairie wheat is shaped by human processes. This is not to suggest that the biology of wheat is completely determined by social processes; rather, the point is to show how biological processes cannot be separated from the social. Wheat is co-produced through the agronomic, scientific, and ecological practices of farmers, scientists, and plants. These co-productions are thoroughly political and involve value judgements about what is agronomically, socially, and economically useful and desirable. Farmers and scientists have an attachment to wheat because of the way it acts in the fields (among other reasons outlined in this chapter). The behaviour of wheat plants in the fields is, in part, the product of years of manipulation and adaptation of genetic structures and farm practice.

NATIONALISM AND THE CULTURAL POLITICS OF WHEAT

Canada is known the world over as a major exporter of high-quality wheat. This reputation has been reinforced at home by state support for the wheat industry, with special attention to the development of quality characteristics, and by the historical importance of wheat to the overall economy of the nation. As a staple food, wheat also enjoys a cultural significance to consumers who understand it as a basic of modern, historical, and religious life. Thus, a strong cultural attachment to wheat exists in Canada despite the fact that the crop has begun to lose its importance to the national economy and despite an urbanizing population with weaker connections to agricultural heritage. Those working to stop the introduction of genetic modification in wheat found that, as a culturally significant food, wheat was very easily and widely politicized. The symbolism and meaning of wheat became an important dimension of struggle and a means through which to engage consumers in debate.

Discourses of wheat nationalism abound in the documents and websites produced by the national and prairie provincial governments and in popular culture, including songs and visual art by Canadian artists and festivals celebrating wheat and bread. The reputation of Canada as a "breadbasket" has been eagerly projected to the world, and Canada is internationally recognized for its high quality, protein-rich wheat. Indeed, Agriculture and Agri-Food Canada's latest (2004) profile of the wheat industry indicated that Canada is the largest producer of high-protein milling wheat and still vies with the United States for the world's largest exports of wheat.[39] While not nearly as significant to the overall Canadian economy as it once was, wheat still earns the most foreign exchange of all agricultural products including livestock. In the years 1999–2000 and 2001–2002 wheat exports earned an average of $3.68 billion.[40]

With an increasingly urban population and a declining number of family farms, it can be expected that the cultural importance of wheat to the nation has waned. Yet, wheat still maintains a symbolic importance on two fronts. First, and especially in the prairie provinces, stories about the wheat economy and the back-breaking labour of settlers, who lived in relative isolation without services such as running water and electricity, maintain a prominent position in official cultural histories reproduced by the provinces. For Eisler, this is one of the myths that has produced the sense of belonging and the emotional and psychological bonds of an imagined (or socially constructed) provincial community like Saskatchewan.[41] Of course, this construction of community is also highly exclusionary to groups that were not part of the pioneering history and it erases the violence done to Indigenous peoples who were forced off the land in favour of wheat and settlers. Nevertheless, as an imaginary of a shared past the wheat economy remains fundamental to many aspects of modern prairie society. Indeed, shortly after coming to power in 2007, the Saskatchewan Party proposed scrapping the wheat sheaf as the provincial logo only to renege shortly thereafter because of widespread public outcry.

The wheat economy also places the prairies in Canadian national narratives; through wheat, the prairie provinces and prairie farmers make a claim to national belonging. As a "breadbasket of the world" the prairie provinces and prairie farmers perform as patriotic global citizens. Indeed, programs like the Canadian Foodgrains Bank, where farmers donate their grains as aid to areas of the world suffering from famine and malnutrition draw on and reproduce the breadbasket narrative. The prospect of delivering genetically modified grain for bread to the global community threatens to undermine the wholesomeness of the bread-basket and the patriotic act of giving. The Foodgrains Bank's rooted-ness in Christian faith provides even more symbolic leverage for farmers concerned about performing their global citizenship.

Second, wheat has maintained its cultural importance as a food that is widely consumed in all parts of the country. One interviewee from the Saskatchewan Organic Directorate highlighted the nature of food as an intimate commodity: "food is cultural and emotional, and canola isn't an emotional food, it's a cooking oil. So it's a cooking oil, you can buy it, you don't have to. But milk is something everybody gives to their babies, to their kids. Milk has an *enormous* sort of emotional cultural position, as does wheat for bread." The cultural and emotional character of milk, like wheat, might explain, in part, Canadian opposition to Monsanto's rBST hormone in milk. Strong opposition to rBST coincided with a lack of opposition to the use of growth hormones in beef cattle, which was approved in Canada in the late 1990s. Part of wheat's emotional and cultural appeal is also its connection to religious life and historical and communal traditions such as "breaking the bread." This was a theme that arose frequently in interviews. One participant from the National Farmers Union expressed that the manipulation of the biological organ-ism may also be experienced as a manipulation of religious symbolism: "And we connect [genetic modification] very sincerely and very clearly to the loaf of bread. And in...the seminary we talk about...breaking the bread. Is it GMO bread you're breaking? And sort of integrate, try and open people's minds to the idea that this deep symbolism, this sacred symbolism is also then being manipulated."

Those working to politicize urban consumers around the topic of genetic modification in wheat found that bread was a food with which urbanites easily identified and behind which they easily rallied. An interviewee with the Canadian Biotechnology Action Network, for example, contrasted the cultural connection that urban consumers have for wheat with the lack of significance and knowledge they have of canola: "People didn't know what canola was...they didn't know what it looked like." Wheat is not only a culturally significant food but also ubiquitous in manufactured food products. Thus, if wheat were genetically modified, it would be very difficult for a consumer to avoid eating GM material. As this campaigner for Greenpeace summarized, "bread, or wheat products, are on almost everybody's table. It's something that... the manufacturers of food products knew would be very, very tricky to take the chance on there being a consumer reaction against their products. "Wheat is used *so* pervasively in so many products that people buy on a daily basis...with canola for example it's far more obscure...canola oil or something like that that people aren't even thinking about when they're consuming a certain food in which it's one of the ingredients. When it came to wheat it's just far more in the face of the consumer."

Those campaigning against GM wheat were, thus, able to mobilize a diversity of ways in which wheat is still culturally important to the public, including peoples' identifications as Canadians, prairie folk, and consumers. As I will show next, the populism associated with wheat's political, institutional, biological, and cultural histories does not exist with canola. Instead, canola is associated most strongly with technological progress and scientific innovation.

· · · · · · · · · · · · · · · C A N O L A · · · · · · · · · · · · · ·

FROM MACHINE GREASE TO EDIBLE OIL
Canola is a much more recent companion species of prairie farmers than is wheat. Rather than a frontier crop, canola (formerly rapeseed) was

first grown commercially on the Canadian prairies during the Second World War to replace usual oil imports (both rapeseed and others) from Europe and Asia. Although rapeseed was a regularly-consumed dietary oil in countries such as China, Japan, and India, it was not considered suitable for human consumption in Canada due to its high concentration of erucic acid, which made it particularly high in saturated fat. Instead, rapeseed was employed in Canada for industrial purposes such as machine grease for steam engines and other war machinery. Facing the supply shortage during World War II, it was soon discovered that rapeseed grew quite well in parts of Ontario and the prairies; and in 1943 the Forage Crop Division of the Canada Department of Agriculture imported 18,640 kilograms of the species *B. Napus* from the United States and distributed it to farmers.[42]

The geopolitical context of blocked oil imports led policy makers to perceive Canada's dependence on foreign oilseed for both industrial and dietary purposes as a national weakness. Self-sufficiency in oilseed became a concern of the Canadian Defence Board, and the idea of modifying industrial rapeseed into edible oil emerged as a priority.[43] A scientist at Agriculture and Agri-Food Canada (AAFC) who was a participant in the process of transformation told me that Canada imported over 90 percent of its edible oils up to World War II: "Although [rapeseed] was originally grown as an essential war commodity to keep the steam engines running…for industrial purposes, it had the potential to be Canada's oilseed crop. And so that was also in the bag in the policy area that Canada should have its own edible oil."

Public plant breeders, thus, began working to transform rapeseed, with two main objectives. First, the erucic acid content would have to be diminished if rapeseed oil were going to be consumed widely. Second, glucosinolates would also have to be reduced in order to establish a productive use for the by-product of the oil — the meal. According to the same scientist at AAFC, the presence of glucosinolates suppressed the function of the thyroid in non-ruminant animals such as swine and poultry. Such animals that were fed above 5 to 10 percent rapeseed meal suffered from numerous problems including poor efficiency in weight gain. The inability to make use of the meal severely limited how much rapeseed could be processed into oil.

The breeding process involved significant cooperation between scientists across several different disciplines at numerous institutions, including Agriculture Canada, universities, and the National Research Council's Prairie Regional Laboratory. For example, a new technology named gas-liquid chromatography (GLC) was developed in order to test the oil properties of increasingly smaller amounts of seed. An early breakthrough in this area was made at the Prairie Regional Laboratory in 1957 and was shared with other labs across Canada. This enabled Keith Downey and his team at the Dominion Forage Lab of Agriculture Canada to perfect the technique, and by 1962 they were using GLC to test half-seeds.[44] This meant that the other halves with desirable traits could be successfully grown out and used in breeding.

While the scientific work described above was being spurred on by the objective of national self-sufficiency, the reality in the fields was stagnation. Although rapeseed was relatively well adapted to the environment and production methods of prairie farmers, it took quite some time to get producers to grow it with much enthusiasm. From 1943 to 1949 the area under rapeseed production in Canada increased from 1,300 to 32,400 hectares with the incentive of a price support program. However, production declined sharply thereafter, with only 200 hectares in production in the 1950–1951 crop year when the price support ended.[45] The end of the war induced a sharp decline in demand, especially because of the conversion from steam to diesel engines, and the crop virtually disappeared from farmers' fields. The Prairie Vegetable Oil crushing plant, built in 1945 in Moose Jaw, Saskatchewan, only stayed afloat because of the small amount of rapeseed it had contracted from growers. It was not until the 1955–1956 crop year that the production of rapeseed recovered to its wartime levels. This was due primarily to the enthusiasm of scientists and plant breeders who had just registered the first all-Canadian rape variety and were promoting it to farmers. As the aggressive breeding agendas continued, production increased, and soon (between 1956 and 1968) a domestic oilseed crushing and extraction industry emerged.[46] Still, much of the early and contemporary production was exported as unprocessed seed to countries such as Japan. It was in 1974 that the first "double-zero" (zero erucic acid and zero glucosinolates) variety of rapeseed was registered by Canadian scientists and the crop's name was changed shortly thereafter to canola (derived from *Can*adian *o*il, *low *a*cid).

The development of canola was driven not only by the goal of national self-sufficiency in edible oil but also by a great deal of hope that the crop would help to diversify the prairie-farm economy and cushion the uncertainties associated with volatile wheat markets. From the beginning, canola was viewed as a high-value cash crop, and farmers could generally sell it at a much higher price than the highest grade of wheat (often close to double per bushel or tonne; see Tables 6 and 7). Although canola involved more risk than wheat, a farmer from the Western Canadian Wheat Growers explained to me that the gamble paid off just often enough. When canola prices are at eight dollars a bushel everybody makes money, but when the price drops to near five dollars the cost of inputs severely limits profits. Compared to wheat, canola uses a lot of nitrogen and other expensive chemicals to control flea beetles, diamondback moths, and armyworms, but with greater fluctuations in price the rewards can be more handsome.

For this reason, political organizing around canola never coalesced as it did around wheat. Furthermore, by the time canola was being grown in any significant amount, a political movement away from the welfare state was well underway at regional, national, and international scales.

For example, it was not until 1978 that canola reached 10 percent of all acres seeded to grain on the prairies. This happened during the final stage of a broad political consensus that saw relatively stable funding and supports for agriculture, both provincially and federally, including co-operative enterprises, plant breeding, transportation subsidies, etc. The production of canola continued to gain momentum as provincial and federal governments began to adjust to and promote neoliberal strategies of governance that saw state supports for agriculture slashed (this happened around the same time as widespread cuts in other industries and social services). In this policy context political organizing among farmers revolved around maintaining the status quo and canola became a cash crop that could buffer the economic hardships that accompanied international recessions and declining domestic supports.

THE INDUSTRY: PRIVATE INVESTMENT AND INNOVATION

Although the development of canola was a state project that required massive investments across several public institutions, the canola industry soon became characterized by its high level of private investment and the absence of institutions such as the CWB that hinder the operation of free markets. A national trade association (initially named the Rapeseed Association of Canada and later renamed the Canola Council of Canada [CCC]) was founded in 1967, and producers were invited to participate alongside the many commercial stakeholders that dominate this organization. Tellingly, the CCC's current mission is "to foster a regulatory, policy and business climate based upon innovation, resilience and creation of superior value for a healthier world; allowing the industry to grow 15 million tonnes of market demand and production by 2015."[47] In other words, canola has undergone a fundamental shift from an object of public investment to an industry dominated by private agendas in plant breeding, input supply, processing, and marketing.

In the initial stage of the transformation of rapeseed, scientists at various public institutions collaborated in order to assemble the necessary set

Table 6. Average farm price in the prairie provinces

YEAR	ALL WHEAT (DOLLARS/BUSHEL)	CANOLA (DOLLARS/BUSHEL)
1910	0.78	n/a
1920	1.58	n/a
1930	0.47	n/a
1940	0.57	n/a
1950	1.52	2.5
1955	1.37	1.77
1960	1.57	1.63
1965	1.68	2.41
1970	1.42	2.33
1975	3.62	5.09
1980	5.59	6.38

Source: Statistics Canada, *Table 001-0017—Estimated areas, yield, production, average farm price and total farm value of principal field crops, in imperial units, annual,* CANSIM (database), using E-STAT (distributor).

Table 7. Annual returns for wheat and canola

CROP YEAR	#1 WESTERN RED SPRING WHEAT (DOLLARS/ TONNE)*	#1 CANADA CANOLA SEED (DOLLARS/ TONNE)**
1985–86	160.00	301.79
1990–91	135.00	287.70
1995–96	263.60	432.29
2000–2001	230.06	289.91
2005–2006	230.10	277.10

* Figures based on producer payments
** Figures based on average farm price
Source: Canola seed data was compiled from the cereals and oilseeds review of Statistics Canada. Wheat data was compiled from Canadian Wheat Board information, http://www.cwb.ca/public/en/farmers/payments/historical. Note that farmer payments made by the CWB are not completely commensurable with average farm prices. Farmer payments represent what farmers were actually paid, whereas the average price is not a good indication of the amount actually received by farmers for their canola. Nevertheless, these are the two best indicators of the relative exchange values of the two crops.

of technologies and knowledges needed to produce a new "double-zero" variety of rapeseed. During the first stage of this process (1944–1966) virtually all investment (85 percent) was public, with only 15 percent coming from private sources.[48] Private players had little incentive to invest in research on rapeseed because Plant Breeders' Rights had not been instituted and most of this initial research investigated the basic properties of the plant and was not easily applied to commercial products.

Beginning to see the potential benefits of a growing domestic oilseed industry, and with the results of a first round of public plant breeding and rapeseed promotion, the Rapeseed Association of Canada (RAC) formed in 1966. Its purpose was to pool the resources of all those with a stake in the canola industry in order to invest in its commercial growth. The RAC had two initial objectives. First, the association worked on trade development in order to ensure that producers would have an expanding market for their crops. Second, funding was channelled into research in the areas of plant breeding, animal feed, and improving quality.[49] For example, the RAC provided monetary resources to Agriculture Canada and public universities for their work on double-low varieties, while the National Research Council maintained its control over and coordination of the collective (industry-wide) research agenda.[50] Meanwhile, the RAC worked more actively on public relations and marketing, and with growers to ensure that they were rapidly adopting new varieties. Producers began to form provincial associations under the RAC umbrella in the late sixties to better influence rapeseed policy, agronomy, and production. This was a period during which most of the RAC members were relatively small, domestic players and during which consensus across the wide variety of actors was possible.

Once the first double-zero variety had been registered in 1974, private interest in the canola industry and in canola research began to grow. Private investment was especially spurred on by the 1978 negotiation of a new International Union for the Protection of new Varieties of Plants agreement, and the domestic buzz that Canada would soon adopt Plant Breeders' Rights (PBR).[51] Private companies began to contribute more to the CCC's research funding program, and the balance within the organization began to shift more toward private companies, especially those associated with agricultural inputs and processors, which were beginning to be bought out by international players. The CCC's funding program, in turn, began to have more sway over public research agendas since it was able to contribute more financially. However, the industry as a whole was still working to promote canola oil as a healthy edible oil and to receive the necessary approvals in markets such as the United States. Thus, research and plant breeding remained tightly coordinated between private industry, public institutions, and the CCC.[52]

When Canada adopted PBR in 1990, private investment in canola really took off. PBR enabled the expansion of legal property rights to the practice of plant breeding that had previously been regulated through non-market arrangements and, according to the six public breeders and researchers who were part of my research, had fostered a culture of collaboration across institutions and breeders. As Kloppenburg explains in the case of corn, PBRs were the social solution to breaking open a new arena of profit-making for private business.[53] Once PBR were legislated and made legally enforceable, those aspects of plant breeding that could yield profit could be successfully enclosed for private gain. Growing private investment in breeding and research was also being promoted by federal and provincial governments that faced budget constraints and began pulling funding for agricultural research. In this context, public research dollars became oriented to complementing the work of private firms, essentially financing the work that was not perceived as profitable. Expressing his frustration in an interview with me, a public plant breeder underscored that this has often meant that public dollars are used in the early stages of research and subsequently these efforts are turned over to private firms for commercialization. Indeed, from 1990 to 1998 only 39 percent of the total investment in canola research was made by the public sector (recall this number was 85 percent from 1944–1966) and 59 percent came from the private sector (recall 15 percent from 1944–1966). More tellingly, the private sector took ownership of 86 percent of the varieties, and 80 percent of the technologies, resulting from this research.[54]

While consolidation and privatization have been characteristic of the post-1990 canola industry, this has not meant that public institutions have been totally shut out. Rather, as Peter W.B. Phillips shows, private seed and breeding companies rely on the knowledge and "basic" research performed at public institutions, and they regularly enter into collaborations or joint ventures with them. In particular, Phillips argues that the development of "know-why" knowledge (which he seems to understand as early-stage basic inquiry) is undertaken almost exclusively by the public sector since this research has not yet been applied to a commercial product. Furthermore, public institutions, because of their long history in breeding and research are vast repositories of "know-who" and "know-how" knowledge.[55] Phillips describes these types of knowledge as "the non-codified knowledge that holds things together" and as involving human relationships and outcomes that are not possible to patent.[56] Private agro-chemical or breeding firms have often been willing to provide minimal profit-sharing for access to this knowledge. Nevertheless, public-sector roles in plant breeding and research have shifted away from proprietorship and the setting of research agendas toward supporting private industry and its agendas. This shift has taken place within the context of private concentration and coordination of research supply chains and the appropriation of extension services by private companies as part of their marketing divisions.[57]

The story of plant breeding and research in canola illuminates some important differences not only between canola and wheat but also between canola and other crops that have received much private investment and attention—for example, corn in the United States. For Kloppenburg, the socio-legal establishment of property rights was just one of two main processes that forced open corn breeding and research for private profit-making. A technological solution that involved scientific manipulation was additionally necessary to overcome the biological reproducibility of corn seed. The development of hybrid corn seed allowed for a technical challenge to the farm practice of seed saving and replanting. Hybrid plants were bred to yield sterile seeds so that farmers would have no choice but to enter the market every year for their seed.[58] Interestingly, as I showed in my earlier discussion about wheat, seed saving remains a culturally significant practice in wheat and has contributed to the lack of private investment in breeding and research. Grown by many of the same farmers who save their wheat seed, canola seed, by contrast, is regularly bought by prairie farmers. According to a scientist at AAFC this is precisely the reason that a plethora of new canola varieties are produced each year. Given that canola is riskier to grow because of its sensitivity to loss and injury, it is easier for the canola industry to market new varieties.

The development of hybrid varieties in canola has certainly furthered the practice of buying, rather than saving, seed. The first hybrid canola variety became available in 1989, and in 2006 an interview participant from the Canola Council of Canada estimated around 55 percent of the Canadian seed market was hybrid; this percentage is likely growing. However, Canadian canola producers seem to be willing to enter seed markets even when their seed is still biologically able to reproduce. Canola boosters have told me that this is due to rapid innovation in canola breeding such that each new year brings several new varieties that promise to outperform the last in areas such as yield, weed control, and uniformity. However, there is more than just innovation that drives seed buying. The agronomic and biological characteristics of canola in the fields also contribute to the culture of seed buying and to a human-environment relationship that is quite different than the one I sketched for wheat.

IN THE FIELDS

Canola has certain biological and agronomic characteristics that have made it a great commercial success and a crop that is particularly amenable to private investment. When contrasted with wheat, the biological and agronomic properties of canola are perhaps more transparently co-produced by the farmers, scientists, and organisms that are involved in its breeding and cultivation. In other words, canola is more popularly conceived as the product of scientific innovation and is more intensively managed by farmers than wheat. Specifically, the crop's weed- and disease-prone disposition in farmers' fields has meant that farmers are very interested in new technologies (including production systems such as "zero till" and genetic modification) and chemicals for weed management. Because of these agronomic difficulties, canola is only feasibly grown in a four-year rotation, and thus canola is less amenable to the practice of seed saving. Finally, and with particular relevance to the case made by opponents of genetic modification, canola is relatively promiscuous and will outcross with other canola plants and with wild relatives.

The variety of canola grown most extensively on the Canadian prairies, *B. napus*, is a cool-season crop that is well adapted to a variety of soils. However, canola is a much more difficult crop to grow than wheat. First, it is not as drought tolerant as most cereal crops; more importantly, canola is highly affected by weeds, pests, and disease. For example, since canola is relatively slow growing and slow to cover the ground, it faces much competition from its many wild relatives (such as mustard and peppergrass) in its early stage of growth.[59] Insects such as the flea beetle and the root maggot also challenge the production of canola; and controlling such pests requires care so as not to damage honeybees or other beneficial pollinating insects. Finally, canola is susceptible to serious disease, especially in areas of intense production and under poor management practices. For these reasons, canola producers have been much more willing to pay for technologies and products that will reduce the risk of loss than they have been for a crop like wheat. Canola has been particularly amenable to private investment and profit-making.

Private involvement in plant breeding and research has in turn shaped the biological and agronomic character and the possibilities of canola production. This recursive relationship has been concentrated around finding more effective chemical controls for weeds and disease, and around agronomic practices that encourage the use of inputs. For example, the practice of zero till is now widely used for canola and is touted as a method of soil conservation, since the seed can be "drilled" into the previous year's stubble and residue, eliminating the need to till. Yet this practice increases the susceptibility of the crop to weeds and disease. It is now recommended that canola not be grown more than once every four years on the same fields in order to prevent the buildup of weeds, insects, and disease. Furthermore, the practice of zero till relies on the genetic modification of canola so that it is resistant to herbicide. Rather than a modification that would require less dependence on agricultural inputs (for example, the engineering of drought resistance), herbicide resistance actually reduces the range of management options available to farmers and increases the susceptibility of canola to weeds and disease. According to an interviewee from Agriculture and Agri-Food Canada, this has made the crop even more difficult to imagine growing organically, since organic growers are restricted from using herbicide and therefore produce crops more susceptible to disease and pests. In the case of conventional growers, the four-year rotation has further reduced the incentive to save and replant canola seed.

The introduction of genetic modification in canola has opened up certain aspects of its biology and agronomic practice to political concern and contention. These concerns surfaced especially during the contestation that surrounded Monsanto's attempts to introduce Roundup Ready wheat in Canada in the early 2000s. By this time, farmers were experiencing unanticipated headaches dealing with herbicide-tolerant canola varieties in their rotations, and scientific studies showed that it would be virtually impossible to grow canola on the prairies without contamination from herbicide-tolerant varieties.[60] Such difficulties were the result of the biological properties of canola, such as its promiscuous nature in the fields, and the policies of importing countries not to accept GM material. Suddenly, the rate of outcrossing in canola (which is the same in GM and non-GM varieties) and the spread of GM material across the prairies became political and practical concerns for farmers, regulators, and the canola industry as a whole. According to a representative of AAFC, canola's promiscuous nature was even more worrisome than other highly outcrossing crops like corn since it is sown on more hectares—5 million, compared to only half a million in the case of corn.

The genetic modification of herbicide resistance in canola concerned farmers not only because of the flow of GM genes across the prairies but also because this canola was difficult to manage in rotation. Seed that had shattered before harvest could remain in the soil for several years and then emerge as a (volunteer) weed in other crops. These plants would then be difficult to control because of their resistance to widely used herbicides. The experiences farmers have had with GM canola, thus, politicized its biology in new ways. This politicization was somewhat unexpected given the cultural legitimacy of canola as a product of scientific manipulation.

NATIONALISM AND THE CULTURAL POLITICS OF CANOLA

The development of canola through extensive collaboration across disciplines and institutions has been a source of pride for Canadian scientists. In fact, in 1982, Canadian public-sector institutions were responsible for all of the six cultivars actively being grown in the world.[61] Despite the fact that the private sector has subsequently taken over the process of registering and commercializing new canola varieties, Canadian scientists and the Canadian state still understand their involvement in canola research as an arena of competitive advantage vis-à-vis other crops and countries, and as a great example of Canadian innovation. In fact, innovation pervades the discourse around canola both in the labs and in farming communities, where canola growers are commonly seen as innovative, forward-looking, and entrepreneurial. As a food, however, canola is not very culturally important to eaters. It carries none of the religious and historical significance that is associated with wheat.

For Canadian scientists, growers, and agricultural policy makers there is a lot at stake in the continued success of canola. Canada has its reputation as the founder and leading scientific innovator of the crop to protect. This was a theme that I heard time and again during the interviews I conducted with scientists, the Canola Council of Canada, Saskatchewan Agriculture and Food, farmers, and even regulators at the Canadian Food Inspection Agency. With the increasing prominence of private investment in canola research and breeding, keeping internationally competitive in canola has meant supporting research agendas and directions that will derive profit. Monsanto's decision to divest from all wheat after it faced resistance to its genetically modified variety (thus compromising a major source of current and future investment in wheat) was often used by interviewees as an example of why public policy, science, and producers should not discriminate between GM and non-GM innovations. This was especially true in canola because, as an interviewee at the Canola Council of Canada said, any rejection of new innovations, or even a slow-down of acceptance of new traits and products, might cause Canadian farmers, state scientists, and regulators to lose the favour of the agricultural corporations. The identity of canola as a product of innovation was so strong among the participants in this research that one scientist stressed that a certain level of arrogance exists around the crop such that the scientific community understands canola as its "baby" and, thus, fully under its direction and care. One researcher at the University of Saskatchewan told me that because canola was brought into existence locally, through scientific labour, scientists saw no need to secure public legitimacy for subsequent innovations such as genetic modification.

Innovation in the canola industry is now understood as one of Canada's competitive advantages internationally. I interviewed an employee at Saskatchewan Agriculture and Food who noted that in the province of Saskatchewan, with its peripheral history and economy, policy-makers have been eager to support agricultural research that might compete with programs in the U.S. and Europe. In Saskatchewan this has meant supporting canola research and genetic modification by mobilizing the discourse of international competitiveness, in particular keeping pace with research investments in the U.S., Australia, and Europe.

The canola industry's embrace of genetic modification is touted as evidence of its forward-looking attitude. Interestingly, this discourse also applies to the farmers that grow canola. They have adopted new technologies such as genetic modification and new farm practices such as zero till that make them innovators and risk takers. Moreover, they must market their crop independently, without the help of institutions like the Wheat Board. According to the discourse of competition, this lends them an entrepreneurial spirit. A representative of the Canola Council of Canada cited the "entrepreneurial and innovative spirit" of Canola farmers as what "keeps guys going out there to grind away in the field." In an interview he linked the entrepreneurial spirit of farmers to the forward-looking, rapidly-innovating nature of the industry, including the adoption of technologies and traits such as herbicide tolerance, hybrid varieties, and genetic modification.

While a strong cultural nationalism associated with innovation exists within the canola industry (including scientists and farmers) a popular culture of canola as a food is relatively weak. I do not wish to suggest here that a popular culture of food is necessarily reliant on nationalist discourses. Rather, consumers attach little meaning at all to canola as food. If anything, eaters identify canola as a Canadian product and as a healthy oil because of aggressive marketing campaigns that had to transform it from machine grease to edible oil. For example, an interviewee who had advocated against genetic modification in canola found that canola lacked the cultural identity that could galvanize a popular campaign of opposition to genetic modification. According to this member of the Saskatchewan Organic Directorate any popular profile that canola might possess is the result of a marketing campaign. Awareness of canola did not come from a deep cultural place; it is a relatively new crop, and consumers can easily replace canola with other oils.

Canola and wheat are, thus, differently produced through nationalist and cultural discourses that do distinct work. The discourses of innovation and competitive advantage that are reproduced in canola contribute to the fashioning of farmers, agricultural policy, and scientific work, but provide a weak ground through which to galvanize consumers. On the other hand, discourses of heritage and spirituality that are reproduced through wheat have the capacity to animate consumers. In the case of Monsanto's RR wheat, cultural and nationalist discourses associated with wheat translated into a strong articulation of opposition and a robust sense of identity as prairie farmers and Canadian consumers.

CONCLUSION

Wheat and canola can both be understood as companion species of prairie farmers, but in radically different ways. This is to say that people and these crops have co-evolved (or as Haraway would put it, people and canola and people and wheat are co-constituted). While it is true that the biologies of wheat and canola have been thoroughly altered by human manipulation and science, it is also true that farmers have adapted their practices and politics to the behaviour of the two crops. Furthermore, the symbolism and meaning of wheat and canola have, in part, made eaters and prairie people who they are.

Both wheat and canola have particular and real materialities that constrain and enable certain practices and relationships, yet this materiality is not fixed, and it is certainly not apolitical. Instead, the particular biological difficulties associated with the production of canola have been harnessed as opportunities for private profit-making; and the production systems accompanying new varieties and technologies have further weakened producers' abilities to re-use canola seed or to grow the crop organically. This contrasts strongly with the case of wheat, which has seen relatively little private investment and is particularly important to organic growers. Wheat, in comparison to canola, is easily managed by farmers and its seed is customarily reproduced outside of the market.

The two crops also have diverging political and institutional histories, which have shaped the characters of the two industries and influenced the meanings of the two crops for producers, scientists, industry players, and the consuming public. For example, the struggle for pools and regulations in wheat have meant that producer-centred organizations like the CWB still hold much clout in the wheat industry and are able to influence breeding and research agendas. The canola industry, on the contrary, is characterized by private investment and an industry organization, the CCC, that must weigh the interests of producers with other (more powerful) players in the industry. For the Canadian scientific and agricultural communities, canola is a powerful symbol of Canadian innovation and competitive advantage. It is a relatively new crop that is understood as a product of scientific manipulation. This contrasts strongly with the cultural politics that surround wheat. Wheat has historical and religious symbolism for Canadian eaters and for farmers, whose agricultural heritage is strongly associated with the crop.

Farmers

MAKE THEIR CASE AGAINST GM WHEAT
Articulating the Politics of Production through Discourses of Consumption

CHAPTER FOUR

In July 2001, while Monsanto was conducting the necessary field tri-als for its application for unconfined release of Roundup Ready wheat in Canada, a coalition of farm, consumer, health, environmental, and industry organizations came forward and announced the varied reasons for their opposition at a press conference in Winnipeg. While six of the nine organizations involved in the coalition were rural and farm orga-nizations, the coalition successfully articulated a broad set of concerns and claims and constructed a diverse and unusual alliance between ac-tors. For example, farm groups actively coordinated with Greenpeace activists and other urban and consumer NGOs that have historically been understood by farmers as radical environmentalists, "crop-pullers" (as one employee at the Council of Canadians described them), and urbanites incapable of understanding the realities of farm life. More-over, the struggle over RR wheat in Canada has brought together more traditional concerns (such as the extraction of profits and control from farmers) with issues such as democratic process, consumer knowledge, environmental and health risks, and much more. In many ways, these were issues behind which it was possible for all the groups involved to rally. All groups in the coalition could agree about the lack of transparency and democracy in Canadian biotech policy and regulation, and the unresponsiveness of government departments and organizations to their complaints.

This chapter looks at the ways in which producer concerns (including both practical attention to agronomic viability and access to markets, and more longstanding questions about how to keep profit and control on the farm) became articulated with and through issues and discourses that are often characterized as consumer-driven. For example, the refusal of Europe and Japan to accept GM material in their food imports became the strongest argument against the introduction of RR wheat, one that farmers advanced by reciting claims about the supremacy of the consumer. Furthermore, criticisms about the lack of democracy and transparency within the regulatory apparatus and Canadian biotech policy were used by all groups to point to the ways that corporate interests had come to outweigh issues such as health and environmental well-being, and public control over the seed and food systems. In this way, the movement against RR wheat on the Canadian prairies provides an interesting contrast with other anti-GM movements in the Global North that have been described most prominently as consumer rejections of Frankenfoods. Such readings, at least in this empirical case, erase the ways in which producer questions are still central to political struggles over GMOs.

PRODUCING AND CONSUMING OPPOSITION TO GMOs

The issue of genetic modification has provided fertile ground for academics in the past few decades, especially since it has erupted as a hotly contested political issue in diverse geographical contexts, across multiple scales, and animating a number of issue areas, from the environment and food safety to the corporate control of seeds and agriculture and much more. On top of the proliferation of academic interest in the regulation of GM foods at various scales[1] there is now a growing body of work that examines social resistance to GMOs.[2] Much of this work focuses on movements in the Global North where resistance has achieved a level of success, for example, by pressuring sub-national, national, and European levels of authority to implement moratoria on the production and/or importation of GM crops and foods.

Of particular note in academic writing about anti-GM movements in the Global North is the analytical attention given by authors to market-based action and consumer fears about food and environmental safety. For example, sociologist Rachel Schurman examined the successes of the European anti-GM movement's attacks on big food retailers, supermarket chains, and Monsanto.[3] While this movement cannot be understood only as a consumer movement (since its main proponents were parties and organizations such as the German Green Party, Greenpeace-Switzerland, the UK Green Alliance, the UK Genetics Forum, and the Intermediate Technology Development Group in the UK), it was directed at consumers and mobilized primarily consumer fears about potential environmental and health risks associated with GM foods. In fact, in another essay that traces the development of the anti-GM movement more broadly, Rachel Schurman and William Munro highlight the fact that casting GMOs as an issue of consumer rights involving risks to health, ethics, culture, and the environment was a particularly successful achievement of European activists.[4] In Scotland, professors Clare Hall and Dominic Moran from the Scottish Agricultural College also found that anti-GM activists' conceptions of the risks posed by GM crops were primarily related to the environment and human health, even though there was some variation across gender and rural/urban lines.[5] Rural sociologist Frederick Buttel agrees that discourses of environmental risk have been at the centre of anti-GM movements, but unlike Shurman and Munro, he argues that this framing is a strategic mistake since it sidesteps issues like the corporate control of agriculture.[6]

Even when academics have set out to study issues that engage with producer/rural politics such as the patenting of life forms or the crisis of natural and agricultural biodiversity, they have often found that anti-GM movements frame their claims and actions in consumer discourses of environmental and health risk. For example, Derrick Purdue, researcher at the Cities Research Centre at the University of West of England, has argued that despite the leadership of agriculture/rural-focussed organizations in Britain, the most common framing of the problem of GMOs has been around environmental risk.[7] Furthermore, one of the most important effects of the anti-GM campaign was to renew environmental movements in the Global North.[8] Where Northern NGOs did advocate for agricultural and seed rights, they focussed their solidarity with peasants and peasant organizations in the Global South, rather than with domestic agricultural/rural concerns.

In North America, scholars have similarly pointed to the prominence of discourses of consumption and consumer rights. This has been the case even when "food and agriculture groups" were included in the analysis. For example, Ann Reisner, professor of agricultural and environmental communications, considered food and agriculture groups including "organic consumers' associations, sustainable agriculture advocates, and consumer advocate groups" as a category of resistance in her analysis of the fight against GMOs in the U.S.[9] Reisner characterized these groups as being concerned, first and foremost, with the unknowable health risks to consumers. Geographer Robin Roff, also focussing on the U.S. movement, affirms that anti-GMO activist groups have framed their opposition in terms of "consumerist" discourses and have chosen to pursue individualist consumer action as their primary method of resistance. However, Roff criticizes such tactics arguing that they constitute a neoliberalization of activism, where the free market is taken to be the most efficient and rational means of optimizing human well-being and where individuals understand themselves only as consumers.[10]

What can be made of the analytical focus of scholars on the politics of consumption, including health and environmental risk, with regard to anti-GM movements? Is it the case that producers have not been able to intervene adequately with their concerns in debates about genetic modification? Have producer interests such as the ability to save seed and maintain farm profitability and control not been attractive to media and scholars? Or alternatively, have farm organizations felt it necessary to articulate their grievances by mobilizing and appealing to consumers and consumption?

The above questions are particularly enigmatic given that a growing literature exists about producer attitudes to GMOs, the incentives to adopt GM crops, and the economic costs and benefits of adoption.[11] Producers are thus recognized as actors that can influence the commercial success of GM crops, yet they figure only marginally in accounts of political opposition. Notable exceptions include the work of sociologist André Magnan and of anthropologist Birgit Muller, who focus on the same producer-led movement at the centre of this research.[12] Social ecologist Chaia Heller provides a last exception with her account of the role of the Confederation Paysanne, a well-known union of small farmers, in the French debate over GMOs.[13] Heller shows how French farmers were successfully able to engage and mobilize the discourse of quality, which is particularly tied to post-industrial agriculture, in order to cast GM foods as antithetical to French culture. In this account, farmers are involved in shaping popular discourse around GMOs and do so by articulating their concerns in terms that mobilize public interest in the preservation of French food culture.

Heller's attention to "quality agriculture discourse" in Western Europe points to an empirical trend that may also be partially responsible for scholarly attention to discourses of consumption and consumer action in movements against GMOs. In fact, geographers studying food and agriculture suggest that consumers (especially in the Global North) are demanding foods with qualities such as local, organic, ecological, and ethical because of the failures of industrial, productivist food systems to provide food safety, environmental sustainability, and public accountability.[14] These empirical turns to "quality" production (that include the increasing prominence and/or resurgence of phenomena such as farmers' markets, fair trade networks, organic production, slow food movements, and various labelling initiatives)[15] are shaping and being shaped by broader economic shifts away from regimes of accumulation (or ways of organizing the economy and work) based on mass-production and consumption. A post-productivist re-orientation in European and North-American agriculture started in the mid-1980s and has promoted a shift in production from quantity to quality, growing environmental regulation of agriculture, movement to more sustainable farming practices, increasing international competition, and declining state supports.[16] In this context, alternative food networks and quality production offer new strategies for development in rural[17] and more recently urban[18] areas that have suffered from the effects of the abandonment of Keynesian organization of production across space and increased competition from trade liberalization.[19] This empirical turn to quality food has opened up space for a politics of consumption that producers can exploit (and indeed have exploited).

The analytical prominence of consumer concerns in anti-GM movements might also be explained by what geographers David Goodman and Melanie DuPuis identify as increasing academic interest in the politics of consumption and the influence of actors that are closer to the consumption end of commodity chains (such as consumers and retailers). They explain that this expansion of interest to the realm of consumption is partly the result of a broader "cultural turn" in social theory.[20] For example, authors such as Ian Cook and Philip Crang assert the potential of consumers as knowing and capable agents who might be involved in resisting the images constructed around commodities through strategies such as "radical passivity."[21] In this way, commodity fetishes (which have been understood as veils concealing underlying social and ecological relations) are recast as political surfaces around which a wide range of struggles might take place.[22] Geographer Suzanne Freidberg's analysis, for example, examines how NGOs and popular media forced British supermarkets into "ethical" reforms of their global supply chains through media-savvy consumer campaigns.[23]

There is thus good reason that little has been written about producers' engagements in anti-GM struggles. An empirical shift to quality production has been demanded by consumers who are disgruntled with the impersonal quality of productivist agriculture and made uneasy by recent food scares. In this context, consumers have been highly responsive to activist claims about the potential health and environmental risks associated with GM foods. Furthermore, the "cultural turn" has heightened academic awareness of the importance of popular culture and the politics of consumption. However, these empirical and theoretical trends do not mean that producer interests and concerns are of little importance to anti-GM movements. Instead, in the Canadian debate around RR wheat, producers were successful in mobilizing their concerns and interests through and alongside a politics and discourse of consumption. New opportunities have been opened up for producers to articulate their vulnerable positions in the political economy of agriculture through discourses and grievances that have wider appeal and the potential to engage consumers in the politics of agriculture.

There was a huge news conference in Winnipeg where we
had this table and we had Greenpeace and the Farmers
Union and the Wheat Board and on and on, and I think
it got sort of neck-snapping attention from government
and Monsanto and everybody else because they were really
surprised by the diversity of the resistance to this stuff.
– *representative of the National Farmers Union*

Well, it wasn't just to take the position...we already had
the position, as did the wheat board. So joining the group,
though, was, how can we sit down at the table with Greenpeace and the Council of Canadians, for example?
– *representative of the Keystone Agricultural Producers*

As the above quotes indicate, working in coalition against the introduction of RR wheat in Canada was both highly effective and tenuous
for the groups involved. As the members of this nine-group coalition
emphasized to me, their strength came largely from their multifaceted
and comprehensive attack on the logic behind RR wheat, including
criticisms that focussed on implications for the environment, health,
consumers, agronomics, markets, democracy, transparency, and corporate power. At the 2001 press conference, where the coalition first announced itself publicly, each group stressed what it knew best from a domain in which it already had some legitimacy. For example, Greenpeace
spoke about the environment, farm organizations about agronomics,
the Canadian Wheat Board about the potential loss of markets, and the
Council of Canadians about consumers not wanting to eat GM foods.
Despite the strength that the coalition drew from its breadth, its co-
hesion was precarious. Members of the mainstream farm groups were

highly sceptical of (being seen to be) working with consumer and environmental groups like Greenpeace and the Council of Canadians. They understood these organizations as potentially threatening to conventional agriculture[24] and disliked their tactics of protest. As one farmer representing the Agricultural Producers Association of Saskatchewan (APAS), whose sentiments I found echoed in many interviews, described, "if they [Greenpeace] would grandstand that press release that we were having down in Winnipeg, we would pull out—meaning [if] some stupid Greenpeacer was climbing a tower or something like that you know we were out." Farm groups were, however, reluctantly ready to set aside their differences with urban environmental and health activists because they recognized that the issue of GM wheat was not only a farm issue and they could not stop it alone.

It was rare for farm groups to work with environmental and health activists, but cleavages across farm organizations also came into play. For example, the NFU was understood as having a more radical position on GMOs, one from which general farm organizations wanted to distinguish themselves. However, despite the NFU's consistent structural criticisms of many aspects of farm policy and reality, it also had a long-established reputation of cooperating and entering into issue-specific strategic alliances with other groups. Farm groups were familiar with the NFU, and its more radical stance against GMOs was accommodated by the coalition. More problematic for many of the mainstream farm organizations was the Saskatchewan Organic Directorate's (SOD) presence in the coalition. With the rapid growth of organic agriculture on the prairies in the previous decade came a certain degree of hostility from so-called conventional farmers who were being asked by their organic neighbours to, for example, modify their spraying practices so that neighbours' fields would not be contaminated. The SOD's categorical rejection of all GMOs was seen among some in the conventional farming community as an insensitive position.

While the groups each approached the topic, at least initially, from different perspectives and strategically used the legitimacy they had already secured in their separate fields of expertise, a few key issues came to dominate the discourse of opposition to GM wheat. Despite struggles over internal dynamics, there was a great degree of consensus about many of the grounds for the coalition's opposition. Most convincingly, the groups argued that RR wheat should not be introduced in Canada because it would threaten existing export wheat markets. This claim was rendered particularly legitimate and forceful because of the active engagement and mobilization of the Canadian Wheat Board (CWB), Western Canada's single-desk marketing agency that had a monopoly in the export of wheat and barley. The CWB was able to gather information quickly from its buyers and quite early on in the debate announced that over two-thirds of its customers would have reservations about buying Canadian wheat if Canada were also growing RR varieties.[25] As a representative of the CWB told me in an interview, concerns about market acceptance were sparked in the late 1990s by customers who, amid introductions of GM soy and canola, began to question whether other crops were also being modified. When the CWB looked into the research being done on wheat and discovered that Roundup Ready was in the pipeline they consulted farmers through focus groups in order to establish a position on GM wheat. This same representative of the CWB told me in an interview that "[what] we heard mostly was that yeah, maybe if there was a disease resistance they might be interested, but Roundup Ready wheat was not something that was required." Knowing that both producers and customers did not want the introduction of GM wheat, the CWB then felt confident about participating actively in the coalition.

In December 2001, the CWB convened a group of farmer, grain industry, technology developer, customer, and federal government representatives (from, for example, Agriculture and Agri-Food Canada and the CFIA) to form the Canadian Grain Industry Working Group on Genetically Modified Wheat. This group was not entirely against genetic modification; rather, it produced a document listing the necessary conditions for the introduction of GM wheat, which included appropriate market acceptance, adequate segregation systems that would keep GM and non-GM wheat separate, more information on how potential agronomic challenges would be dealt with, and a positive cost-benefit ratio for farmers.

It was not only the grain industry and farmers who placed so much importance on the market argument. The Council of Canadians (CoC), the Canadian Health Coalition (CHC), and Greenpeace Canada also emphasized that consumer polling in Canada showed that the majority of Canadians would choose not to eat GM products if they could distinguish them from non-GM products. The CoC began their testimony to the Senate Standing Committee on Agriculture and Forestry in November 2001 on precisely this point: "Genetically engineered foods were introduced into our food supply without our knowledge or consent. Today, they account for up to 70 per cent of processed foods found in our grocery stores. Though Canadians have expressed clear concerns over the lack of proper testing of these foods on public and environmental health, this government seems determined to continue releasing new GE products, such as GE wheat, into our food supply. Poll after poll[26] has shown the growing unease and bafflement at the fact that we are being forced to consume foods that could potentially be harmful to our health."[27] Being an international organization with coordinated campaigns across several countries, Greenpeace mobilized both domestic and international market opposition to GMOs and placed the market non-acceptance of GMOs at the forefront of its campaign. In fact, the market argument became so strong that *all* groups involved in the coalition used it very frequently and prominently in their policy statements, press releases, and public discourse.

A second argument that articulated both producer and consumer concerns centred on the potential environmental impacts of unconfined release. Here the coalition was concerned with the possibility of completely unknown and unknowable environmental risks, plus a set of more familiar and predictable impacts that were based on experience with RR canola. The general farm organizations Agriculture Producers Association of Saskatchewan (APAS) and Keystone Agricultural Producers (KAP) were particularly keen on showing how the introduction of RR wheat could threaten the environmental benefits associated with the practice of "zero till" agriculture that has been adopted quite widely on the prairies. Ironically, it was the introduction of RR canola that spurred on the practice of zero till in which farmers "drill" their seed into the ground through last year's stubble. Instead of tilling to knock down weeds, Roundup is applied to "burn off" weeds and volunteers from previous rotations. A second Roundup Ready crop in rotations, they argued, would threaten the viability of the practice of zero till because volunteers would be Roundup-resistant and would therefore require the application of a second herbicide or the need to till. A member of APAS reinforced in an interview that "We conserve moisture with zero till, we reduce fossil fuel consumption, we improve soil quality, we reduce soil erosion. Now I don't know if you're old enough to remember, when we used to drive down the roads in this province, the dust storms we had...You know the whole bloody province would have looked like the dirty thirties, eh?...We're saving the soil with zero till."

Not surprisingly, Greenpeace also highlighted ecological disruption as one of their main concerns with RR wheat. They echoed many of the ecological concerns raised by the various farm groups and added that RR wheat could have deleterious effects on biodiversity and on soil biota. Specifically, Greenpeace was concerned with the consequences of increased reliance on glyphosate (the generic name for the active herbicide in Roundup) that a second Roundup Ready crop would engender. Biodiversity would be negatively impacted by the increased use of a herbicide that kills everything other than RR crops, and although Roundup had been scientifically shown to have little toxic effect on animals, Greenpeace was concerned about its effect on fish and marine ecosystems through runoff.[28] Two studies on the effects of glyphosate on soil biota also allowed Greenpeace to argue that glyphosate disrupts soil organisms in a way that increases agronomic challenges on the farm.[29]

The Saskatchewan Organic Directorate (SOD) was also key in spreading the discourse of environmental harm during the debate over RR wheat. Being committed to farming practices that exclude agricultural herbicides like Roundup, the SOD made strong arguments about the trait selected for modification being Roundup Resistance. In particular, the directorate attracted a lot of media attention because of their attempts at the certification of a class action against Monsanto and Bayer (the latter is responsible for a different GM herbicide-tolerant canola called Liberty Link) for the de facto loss of canola as an organic crop. Here, contamination through pollen flow or wind (especially in winter when seed can blow long distances across snow-covered fields) has meant that the production of organic canola in most parts of the prairies is no longer possible. On top of the risks associated with pollen flow and blowing seed, it is nearly impossible to get GM-free seed. Thus, the SOD was able to draw attention to the threat that the genetic modification of field crops (and particularly of wheat because of its prominent position in most organic and non-organic farmers' rotations) poses for the viability of organic farming on the prairies. In a new book about the politics of GM science, sociologist Abby Kinchy provides an in-depth analysis of the SOD's court case against Monsanto and Bayer Crop-science.[30] She shows how these claims about environmental impact were tied up with arguments about lost export markets.

The third argument that came to dominate the discourse of opposition to GM wheat revolved around the lack of democratic and transparent process in the development of biotech policy and regulation in Canada. The many arguments that I heard and unearthed during my research that fit into this category are simply too many to summarize here, but the few that I do address were the most prominent. In their policy documents and presentation to the House Standing Committee on Agriculture and Agri-Food in 2003, the Saskatchewan Association of Rural Municipalities (SARM) listed the secrecy of the field-trial locations of RR wheat as their number two concern, following only the possible loss of markets. By 2003 the CFIA had denied farmers' requests to know the locations of the forty-five Monsanto field trials across three provinces.[31] They were worried that their own fields might become contaminated with GM material if they were too close to the secret test sites. Because of their experience with GM canola, which had, by the early 2000s, thoroughly contaminated the canola seed supply and handling system, producers knew that outcrossing and the spread of GM wheat seed by wind were real possibilities. In a 2003 memo to then-Agriculture Minister Lyle Vanclief, Stephen Yarrow, head of the Plant Biosafety Office at CFIA, admitted that "the CFIA is caught between on the one hand wanting to provide information about its activities to Canadians and to protect their interests, and on the other hand protecting the business information of researchers conducting confined trials."[30]

All the organizations involved in the July 2001 coalition highlighted in their press releases, policy documents, testimony to the Standing Committees, or in their interviews with me that no meaningful public debate had preceded the introduction of GMOs in Canada. Major stakeholders had not been consulted on individual modifications, and many even argued that consumers, farmers, and the general public had been purposefully kept in the dark. For example, when I asked about why the release of GM canola was not highly politicized, my participants explained that it was not widely known that it was under approval, and that any information they did have about GM canola was from Monsanto's pre-release promotions. Furthermore, the Council of Canadians, the National Farmers Union, the Canadian Health Coalition, the Saskatchewan Organic Directorate, and Greenpeace Canada all pointed to the failure of the Canadian state to implement a mandatory labelling scheme for food containing GMOs (which had been under consideration and before Parliament but was ultimately denied for vote in 2002). They argued that mandatory labelling was a necessary precondition for transparency and that the government's intention was to silence consumers on the issue of GMOs by taking away their ability to discriminate between GM and non-GM products.

Charges of secrecy and a lack of democracy were also applied to other branches of the Canadian government, including Agriculture and Agri-Food Canada (AAFC), which was a partner in the development of RR wheat[33] and was also involved in research about the ecological, agronomic, and economic impacts of GMOs. In fact, for many of the groups opposing RR wheat, the only way to get answers to their questions about AAFC's relationship with the biotech industry, and even the results of AAFC's scientific research, was through filing requests under Canada's Access to Information and Privacy Act. A member of the Canadian Health Coalition told me in an interview that Greenpeace and the CHC collaborated in filing more than a thousand requests in the early two thousands; representatives from NFU and SOD reported that farm organizations that had heard about an AAFC study of GM contamination in pedigreed seed lots were not provided with the data until they filed a request. In an interview with a scientist at a Canadian public university, the situation was summarized in this way:

> The other unfortunate thing that we've seen in Canada is that, at least [with] the wheat situation, is that Agriculture Canada scientists were given a gag order. I mean even if it had never been written on paper, when you talk to those scientists they have been told that they cannot make public comments on anything related to GM, period. Scientists in public institutions, at least government institutions [have] been told they can't talk about this, even though they could have contributed to the conversation. So what did that leave? That left university people...And even amongst university researchers, if they happen to be under contract to do research for one of these private companies they were put in a position where they couldn't comment. So there were very few people that could really provide a good balanced commentary on anything related to that.

As I will show, issues of democracy and transparency were almost always closely linked to claims that the Canadian state was not acting in the public interest, but instead reorienting research and regulation to favour private commercial interests.

PRODUCTION *THROUGH* CONSUMPTION

> So, if we cannot grow organic wheat, there cannot be organic farms or grain farmers. And if we can't be organic grain farmers then there's no organic food. You lose the food supply, because we've lost the farmers. And I think that was the argument that was the strongest for us. It's our livelihood, it's our way of life, it's our history, it's our future,

it's all bound up in whether we can. If one of our primary crops is not available to us, we basically die.

— *representative of the Saskatchewan Organic Directorate*

Well there's producer interest narrowly and broadly. Narrowly, you know, just, does this stuff really make you more profitable? But broadly...it is a question of who is going to control agriculture...the real struggle was who would control the food system, farmers or corporations? And the pivotal battle in that larger struggle was around seed. And the thing that was in play was wheat. Would wheat become the high-priced patented seed that was sold and controlled by a few corporations or would it remain a seed that was largely farm-saved and provided by farmers and controlled by farmers and reused at no cost?

— *representative of the National Farmers Union*

At first glance, the coalition to stop RR wheat seems to be a good example of a new social movement in a post-industrial agri-food landscape. In the coalition's press releases and policy documents, testimony to the House and Senate Standing Committees, and comments from interviews one can easily identify the increased need to think about consumers and their politics at sites of production, the prominent position of environmental concerns in current discourses of consumption and production, and the tendency to prioritize demands for more democracy over criticisms of capitalism. Yet, concerns over the extraction of surplus (profit) from agriculture and the control of food systems have not been totally erased; rather, they have been strategically articulated through discourses of consumer sovereignty, environmental risks, and democracy. In this way, the movement against GM wheat on the Canadian prairies challenges academic characterizations of anti-GM campaigns as being driven by consumer and environmental politics.

The market-acceptance argument was particularly successful for prairie producers because they could easily appropriate the discourse of "the consumer knows best" from the extension services and recommendations about adapting to the future agrarian economy of various branches of the Canadian and provincial governments. The idea that farmers should spend more time and effort developing niche markets and thinking about the consumers of their products can be seen in the literature on quality food networks that posits a more discriminating, knowledgeable, and demanding consumer in the post-industrial world.[34] This is a discourse that I witnessed in action at the five farm meetings I attended during my field research; it applies even more strongly to organic producers, who are part of an industry that is more consumer-driven than their conventional counterparts. Rather than producing "commodity" wheat for the bulk undifferentiated market, farmers were being instructed to get (or stay) competitive by developing their websites, reaching out to contractors, and growing specialty products that could be tailored to the exact specifications of buyers. Thus, when the CWB made public that over 80 percent of Canada's export markets for wheat would not accept RR varieties, the producer organizations had found their silver bullet. Testifying to the House Standing Committee on Agriculture and Agri-Food in 2003, APAS argued that "farmers in Canada have been encouraged to be more responsive to the consumer and design their production for the marketplace demand. We need to be very sensitive to this demand....It is not reasonable to encourage producers within a country to cater to market demand and then register and release for production a crop that has no consumer acceptance."[35]

Producers mobilized the "consumer knows best" discourse in order to advance a concept of their own interests as precariously positioned vis-à-vis other actors in wheat commodity chains, including powerful corporations such as Monsanto and empowered consumers upon whom they were quite dependent in their successful opposition to RR wheat (but upon whom they could not always rely). For example, a representative from the SOD explained that farmers would have to accept GM wheat if consumers of organic food decided that they could accept some level of GM contamination. According to the interviewee "right now our argument hinges quite a bit on the fact that the European market for organic food is very strong on anti-GMO and if that changes...that really knocks the props out from underneath our argument here. I think it would be really unfortunate because I think it opens up the door for all kinds of loss of control over our food system."

Cloaked in the "consumer knows best" language, all producer groups demanded the addition of a cost-benefit analysis to the regulatory approval process, which reinforced the notion that primary producers have a legitimate and distinct economic stake in wheat commodity chains. Without specific protection, producer interests would be undermined. As a representative of the National Farmers Union said in an interview with me, by the time the argument about losing international wheat markets gained traction farmers had successfully advanced the idea that RR wheat was of no benefit to farmers. By then it was "strictly seen as a corporate benefit...And there was really very little discussion even in the Farmers Union that we should back off, that we shouldn't be so critical, that some of our members wanted this stuff. By then nobody wanted it."

Indeed, the opposition of producer interests to corporate interests can be found in the discourse of all the groups involved, not just those with a history of radical agrarian politics such as the NFU. The SARM, for example, clearly, described this opposition to the House Standing Committee on Agriculture and Agri-Food in 2003: "The reason we are in favour of a regulatory approach versus a volunteer approach is to ensure that farmers' interests are given adequate consideration in the process. A voluntary approach that relied on a technology developer to withhold developments would not give enough weight to farmers' interests. Developers may not have the incentive to withhold products from the market. It's their development. They spend a lot of money. They want to get it out there."[36]

The fact that environmental issues came to occupy a position of significant importance in the debate around RR wheat also, at first glance, seems to position this struggle as characteristic of a consumer-driven anti-GM politics. Interestingly, however, the arguments about environmental risk put forth by the coalition cannot be understood as underlain by a conception of nature as external to social relations,[37] which historian William Cronon has shown underlie the motivations of many conservation agencies.[38] In this case, nature was not primarily understood as a realm separate from humans and social relations, as pristine, and in need of saving. Instead, all the groups involved connected genetic modification to specific agronomic implications and the risk of creating "new" natures with unintended consequences. Greenpeace trod most closely to the "nature as external" conceptualization, arguing, for example, at the Senate Standing Committee on Agriculture and Forestry in November of 2001 that "crop management responses to these problems [referring to the transfer of herbicide resistance to wild plants] such as increased fertilizer use, shortened crop rotation and shifts in pesticide use that increase toxic loads in the environment could have devastating effects on natural soils, terrestrial and aquatic ecosystems. For example, they could cause shifts in food sources and habitat for insects, soil organisms and birds and their predators. Or they could result in contamination of soil and groundwater from pesticides."[39] Yet, in this testimony and in other documents such as *Against the Grain*, Greenpeace discussed the prairies as produced natures and working landscapes.[40] Thus, for Greenpeace, environmental risk factors included the possible impacts of farming on insects, soil organisms, etc., and also questions about what particular natures were agronomically practical and manageable for farmers.

The conception of the prairies as produced natures and working landscapes was an environmental discourse behind which even producers who had felt threatened by ecologists in the past could stand. It was also a discourse that they knew had significant potential for coalition building. Farm organizations could engage consumer groups and the general public by talking about issues like invasive weeds and at the same time get the public to think about the nature of the work that farmers do. Weeds (including the potential increased weediness in wheat or the development of herbicide resistance in other plants through outcrossing or adaptation) became a central environmental concern for all groups involved and pointed to a conception of nature as inseparable from human values, intention and production, though neither completely controlled nor controllable. The production of new organisms was simply too risky because of the organization of on-farm practices such as zero tillage.

The serious lack of democratic and transparent process in the regulation and policy-making around GMOs in Canada was another message that the Canadian media and public found compelling. Demands for more democracy and transparency were methods through which the organizations working in coalition articulated the conflicts of interests that they perceived between a Canadian state charged with both regulating and promoting GMOs and between public needs (in terms of, for example, consumer choice to avoid GM foods and the viability of family-farm production) and for-profit private industry. However, demands for more democracy and transparency were not only strategic methods for mobilizing producer questions but also legitimate claims in and of themselves. Still, such demands most often appeared alongside arguments about who would lose and gain from the introduction of RR wheat, while arguments about the corporate control of food and seed systems were most often made with reference to the failure of the state to properly protect public interests.

Of particular interest to all the organizations involved in opposing RR wheat was the conflicting role of the state, which is responsible on the one hand for the promotion and advancement of GMOs, particularly through Agriculture and Agri-Food Canada (AAFC), and on the other hand, the regulation of GMOs, primarily through the Canadian Food Inspection Agency (CFIA). As many of the groups pointed out, the AAFC and the CFIA report to the same minister and work collaboratively on much of their policy and practice. A representative from the NFU summarized the blurring of lines between regulation and promotion in this way: "many, many people have said that the CFIA sees the corporations as the client and the citizens as somehow a nuisance. And we got that impression that they were there—in the absence of easily identifiable proof of health damage or something like that—they were there to work with the seed companies to move their new GM products through the system."

Not only that, but producers (from every organization that I interviewed) also articulated that the state's retreat from agricultural research through the AAFC, public universities, and at provincial levels has added to the vulnerability of farmers. A Saskatchewan Organic Directorate representative commented,

> I think the real tragedy is it's to the detriment of [the] good-old-fashioned publicly funded research in agriculture that both levels of government used to provide. And it was always seen as a social benefit across the board for all society...to have that publicly funded research. Unfor-

tunately what's happened is that government has decided that they're gonna put taxpayers' money in funding biotech companies, chemical companies. And departments of agriculture at the university level and elsewhere have ended up being turned into avenues for using public money to fund private for-profit enterprise. Public money's going in to line the pockets of the shareholders of companies like Monsanto and Bayer through the university departments of agriculture.

For producers, the outright catering to biotech companies became even clearer when, in 2002, the CFIA removed the market test criteria that an advisory committee, through the Prairie Registration Recommending Committee for Grains, could use when evaluating applications for the registration of new varieties. The market test was a mechanism that had fallen out of use, but that recommending committees were considering mobilizing in order to stave off GM wheat. According to a representative from the SOD, the CFIA was under pressure to change its variety registration process so that the issue of genetic modification could not be brought to recommending committees that included farmer participation. Instead, an advisory committee, perhaps without producer participation, would deal with the issue. According to a SOD interviewee, such an advisory committee would be "more rubber stamping than anything else, and here too a company like Monsanto submits data. CFIA reviews it, they don't do their own independent testing. And based on an evaluation done by a couple of people in an office maybe they're gonna say yes or no, and there hasn't been any nos. Every single genetically modified food crop in Canada...[has been approved]."

CONCLUSION
At first glance, the arguments that came to dominate the discourse of opposition against RR wheat in Canada seem to be characteristic of a

post-industrial politics of food wherein new actors (such as consumers), new issues (such as the environment), and new sites (such as state regulatory apparatuses and scientific laboratories) become politicized. However, in the struggle against RR wheat in Canada, discourses of markets, the environment, and democracy were almost always articulated alongside producer concerns about the profitability of agriculture and the control of food systems. In fact, arguments about market acceptance were used to expose the vulnerability of producers in comparison to other actors in the wheat commodity chain; claims about possible environmental risk were used to advance a notion of nature as produced and the prairies as working landscapes; and complaints about a lack of democratic and transparent process were entangled with charges of conflict of interest between a public sphere that is supposed to protect public (including producer) interests and for-profit corporations attempting to gain more control over food systems.

By mobilizing the discourses of markets, the environment, and democracy, the coalition at the centre of this research was ultimately successful in pressuring the Canadian government and Monsanto to withdraw their application for unconfined release of RR wheat. However, opponents of biotechnology are unlikely to be able to use the same arguments in future biotech battles in Canada. The argument that galvanized the most support during the campaign to prevent the introduction of RR wheat has been significantly weakened as the European Union has slowly approved more genetically modified food and crops (even while many member countries maintain their own bans). For example, the moratorium on GMOs ended in Europe in 2004 and in 2006 Dupont's GM maize was approved for all food uses in the EU. Furthermore, research into modifications that would be more consumer-friendly (such as altered oil profiles in canola) are underway and could undermine consumer opposition to GMOs. In this context, producers will need to find alternate methods of advancing their claims to farm profitability and control.

Demands for more democracy and transparency are other claims that will be hard to sustain in future biotech battles in Canada. This is not because such issues have been resolved by the Canadian government, but because the CFIA has institutionalized mechanisms such as public consultations that provide a veneer of public accountability and transparency. In fact, the Canadian state is quite intent on changing the public perception of GMOs and has begun to fund pro-biotech education campaigns and large biotech research centres in public universities that require partnerships with private industry. Will social movement organizations' claims about the conflicts of interest that characterize the Canadian state's treatment of biotechnology policy and regulation continue to be perceived as legitimate if public consultations and commercially driven research become routine and normalized?

As Frederick Buttel has argued, the "environmentalization" of the arguments put forward by movements against GM crops also present possible political ambiguities for the future.[41] While the Canadian movement against RR wheat mobilized the environment in ways that linked it intimately to the work that farmers do and the negative re-working of agronomic practices required by RR wheat, it is possible that future modifications will provide real environmental benefits. For example, Buttel suggests that most scientists trained in the last two decades see genetic modification as a research method with potential for breeding traits that could contribute to sustainable agronomic practices.[42] In this case, anti-GM activists will want to develop allies in the research community so that they can ensure that research is done not for the private benefit of agro-chemical companies but with goals such as reducing producer risk and inputs such as water and agro-chemicals.

In other words, producer interests, both narrowly and widely conceived in terms of keeping profit and control on the farm rather than in the hands of companies like Monsanto, may not preclude the use of genetic modification *tout court*. Certainly, farmers will want to be driving the conversation about agricultural research agendas (including their environmental implications), whether these agendas include the use of genetic modification or not.

At stake in this debate are the consequences associated with what sociologist Harriet Friedmann has identified as an emerging "third food regime." For Friedmann, this third regime was precipitated by a world food crisis, beginning in 1973, that called into question the organization of international and national relationships associated with the second regime, including heavy state involvement in price setting or direct payments to farmers, import controls, subsidized exports disguised as food aid, and the transnational industrial integration of commodity chains. Responding to the demands of social movements in dealing with these food crises, Friedmann suggests that a corporate-environmental regime is beginning to consolidate. This "third food regime" is led by corporate retailers that reorganize commodity chains to combine environmental politics with increasingly privatized governance systems such as labels and standards (including organic, fair trade, ISO, etc.). [43] On the one hand, national standards in food safety and quality are lowered through international agreements to liberalize trade; while on the other hand a private set of more stringent standards is erected and valorized by wealthier consumers resulting in a polarized food landscape.

While the articulation of producer interests through discourses of markets, the environment, and democracy were sufficient in staving off RR wheat in Canada for the time being, it is clear that the control of genetic modification by large multinationals such as Monsanto has not been sufficiently challenged. Producers and consumers must think hard about which discourses and strategies can successfully push back the corporate environmental takeover of food systems, regulations, and standards. The appropriate strategies and discourses will need to engage and mobilize consumers concerned about the fate of food systems alongside farmers who bring valuable knowledge and an interest in keeping control and profit on the farm. Future battles over GMOs must name the set of social and environmental problems that accompany private for-profit agricultural research and the ways in which regulatory standards, farm organization, and consumer "choice" are being handed over to corporate control.

Only Consumers
GET TO DECIDE IN THE MARKET
Canadian Farmers Fight the Logic of Market Choice in GM Wheat

CHAPTER FIVE

> Modern individuals are not merely "free to choose,"
> but *obliged to be free,* to understand and enact their lives in
> terms of choice. They must interpret their past and dream
> their future as outcomes of choices made or
> choices still to make.
> — *Nikolas Rose, Powers of Freedom: Reframing Political*
> *Thought (Cambridge: Cambridge University Press, 1999), 87.*

Perhaps the greatest challenge for the Canadian coalition against
Roundup-resistant wheat was countering the view that the fate of RR
wheat should be decided in the marketplace through the mechanism
of individual demand. Indeed, strong support for individual consumer
choice in the market was well represented by proponents of GM wheat,
including biotech companies, trade associations and lobby groups, plant
breeders, scientists, regulators, and farm organizations. Often citing the
success of RR canola and the widespread adoption of the technology
among prairie farmers, such advocates insisted that the only impartial
method to decide the future of GM wheat was to introduce it into the
market and let individual producers and consumers choose whether to

buy it based on their own specific needs. The marketplace was here represented as the only appropriate site and mechanism for social change. According to this reasoning, the proper role of the state was to make individual choice the organizing principle of the economy.

Directly in opposition to the idea that the fate of RR wheat should be decided in the market, farmers and consumers (of food) argued for a negotiated and collective decision to ban RR wheat. In advocating against commercialization, opponents of RR wheat advanced notions and examples of collective action and common good. For example, by recalling their history, representatives of farm organizations asserted their capacity to act collectively in order to assert their interests in the production, marketing, and distribution of grain. Canadian consumer organizations, at the same time, reinforced the idea that the public should be able to choose to keep certain products and technologies out of the market for enough time to make definitive conclusions about their safety and environmental effects. In other words, opponents of RR wheat argued that consumers and producers should have the right to decide the fate of RR wheat through collective processes, including public-policy measures.

In what follows, I examine the common-sense notion of the right of consumers and producers to choice in the marketplace by juxtaposing it with the discourses of collective action advanced by anti-GM activists. The two different discourses are based on radically different presumptions about the capacities of individuals to fashion their futures. As I will argue, the first conception of choice wherein consumers (as buyers of food) and producers (as buyers of inputs) individually "vote with their dollar" on the market forecloses options for collective action. For example, it strips farmers of their common interests and experiences as producers of food (since it conceptualizes them only as buyers of inputs) and limits the spectrum of possible political expression to individual acts of consumption. In the realm of food consumption, this first discourse similarly pre-empts action beyond self interest as it suggests that only individual purchasing power can decide the fate of RR wheat.

GOVERNING THROUGH CONSUMPTION

There is no doubt that consumers can change the world in some way; however, it is important to think carefully about the nature of change that is based on individual market action. Social theorists Nikolas Rose and Mitchell Dean have both written extensively about the way in which the concepts of individual choice and freedom underpin modern approaches to governing society. Their insights help to explain what was at stake in the argument that individual consumers should decide the fate of RR wheat by "voting with their dollars" on the market. Rose and Dean draw on the concept of governmentality that is derived from the work of Michel Foucault. For Foucault and those following in his tradition, it is important to think about the act of governing (that is, of directing, managing, controlling, and regulating human conduct) not only as centred in the power of the state and its bureaucracy but also as pursued by a variety of non-state actors and institutions (for example, by markets, health care practitioners, or community organizations) and even as perpetuated by individuals themselves.

For Dean, governmentality "deals with how we think about governing...[and] emphasizes the way in which the thought involved in practices of government is collective and relatively taken for granted."[1] In this respect, current practices of governing society can be understood as involving specific ways of thinking about how to manage the conduct of people. For example, under neoliberal economic relations the discourse of individual market choice becomes a principle or vocabulary through which individuals govern themselves (i.e., individuals regulate their own behaviour, hopes, and desires through the discourse of market choice). Government (managing oneself in relation to market choice) is here understood as an active process that individuals will exercise in their everyday lives. The notion of consumer choice seems to be a diffuse discourse (not rooted in any singular institution or state) that shapes neoliberal subjectivity writ large; it is taken for granted in how individuals and groups act in the world. As Rose suggests, subjects of neoliberalism necessarily understand themselves and their relationships to the world around them as constituted by personal choices and the exercise of freedom.[2]

The idea that personal choice and the exercise of freedom underpin how individuals and organizations act in the world has been taken up by geographer Julie Guthman. According to her, "probably the most central organizing theme in contemporary food politics is consumer choice. That this seems to go without saying suggests the extent to which this notion has become taken for granted."[3] Food activists and scholars alike seem to be quite taken with the possibility that consumers might express their politics and identities by means of their consumption, and through this practice of voting with their dollars, force meaningful social and environmental change in food systems. Yet, by looking at the history of consumer activism, it becomes clear that the notion of voting with one's dollar is a somewhat novel phenomenon that reinforces an individualized and neoliberal practice.

A growing body of social science research on the concept of consumer choice and the relationship of the individual consumer to a broader community or society contributes three main points that are relevant to the conflict over RR wheat. First, consumer activism has not always been confined to the individual practice of "voting" on the market. Instead, consumers have organized in ways that challenge the very logic of the market and the individuality that characterizes current consumer activism. Second, current definitions of consumer choice are narrow, and refer most often to acts of market consumption. While current consumer activism might claim to promote ethical outcomes, any means of acting collectively, especially those that challenge market logic, are out of sight. Third, this narrowly defined conception of economic choice has come to apply universally to widely varying realms, including public service provision, environmental protection, and community action.

UK professors of food policy and management Tim Lang and Yiannis Gabriel break the history of consumer activism in the West into four waves, each with distinct values and methods of organization and collective agency. The first wave identified by Lang and Gabriel began in the early 1800s in Britain and was marked by its working-class character. The co-operative movement sought to supply working-class families with the basic consumer necessities of life at affordable prices that excluded profit. The principle of "self-help by the people" motivated these producers to work together in the realm of consumption, and the goal was to organize outside of regular market imperatives like competition and profit seeking.[4]

Such consumer co-ops were also important in rural North America in the late 1800s and early 1900s and developed alongside the already successful producer cooperatives. Consumer co-ops helped consumerism take root in rural North America, but this consumerism was inflected with struggles about how rural people should modernize, including to what extent and how rural folk should engage with the market.[5] As Ronald Kline argues, the multiple possibilities ushered in through consumer choice allowed farmers to adopt some elements of consumerism while maintaining their independence and fashioning new rural cultures and new forms of rural modernity.[6]

The second movement, named "value-for-money" by Lang and Gabriel, came into fruition in the 1930s, especially in the U.S. This second wave publicized the growing power of food corporations that were increasing their market shares through combination. Organizations like Consumers Research Inc. were founded in order to research product safety and offer information so that consumers could be more effective in their market activity and pursue the best value for their money. In the early 1990s, however, this form of consumer activism started to wane because of the challenges of providing consumer information for niche markets and because of the emergence of multinational retail giants that offered rock-bottom prices.

Lang and Gabriel name the third wave of consumer activism "Naderism" after Ralph Nader—the author, activist, and presidential candidate in the U.S. 2000 election. In America this movement consisted most prominently of a network of organizations assembled by Nader and his co-workers; they sought to build grassroots public pressure for stronger regulations and standards of conduct for corporations of all types from the health sector to automobile manufacturers. All levels of government were called on to protect the individual as a citizen against corporate giants. Unlike the second wave, the idea was not to empower consumers in the market, but rather to constrain and limit the market through state regulation.

The fourth and last wave identified by Lang and Gabriel began slowly in the 1970s and gained strength in the 1980s and 1990s. Now enjoying significant clout, and receiving much attention from academics and activists, "alternative consumerism" addresses a variety of concerns through individual purchasing of green, fair trade, ethical, and organic products. Originating in Europe as part of the "green" movement, alternative consumer activists deployed some of the same strategies as their second-wave counterparts, comparing the consumer products and practices of various companies and thereby encouraging producers to compete for (perceived) environmental soundness. Given earlier concerns in the green movement with reducing consumption, alternative consumption marked a break in environmental activism. Consumerism was no longer something to be curbed; rather, individual purchasing power was to be championed if used in conscious and strategic ways.

The notion of consumer choice that animates this fourth wave of consumer activism is generated and reinforced by contemporary political theorists in the rational choice tradition. According to political scientist John Dryzek, rational and social choice theories[7] are premised on the example of *homo economicus* pursuing his preferences and goals in the marketplace. Here, the choices of individuals behaving strategically and in their own interests can be aggregated to yield the optimal collective decision.[8] The market is thus the most democratic decision-making instrument because of its transparent capacity to aggregate private preferences. In this perspective, the only just form of collectivity is one that has no effect on fully formed expressions of individuality.

Of course the underlying assumptions of rational and social choice theories have been subject to intense criticism by scholars in a variety of fields.[9] Advocating for deliberative democracy, Dryzek criticizes these perspectives for their assumption that interests and preferences are individual and objective expressions of autonomous subjects. Instead, Dryzek understands preferences as socially constructed (the product of social relations, thus reflecting social norms, discourses, and structures) and individuals as being persuaded and persuading others through social and political interaction. This conception of preferences as the product of dialogue and negotiation, and as reflecting social relations, leaves room for the possibility (even necessity) of collective decision making and a public sphere. Furthermore, there exists the possibility that individuals with different perspectives and different social locations might engage in negotiated collective action.

The argument that individuals can be persuaded to understand and act upon (or according to) a common good is fundamental to notions of citizenship (including social, legal, and political conceptions). This is a perspective that sees citizens not only as exercising civil, political, and social rights but also as responsible to carry out the ethical obligations that accompany such rights.[10] What happens, then, when the ethical obligations of citizenship are reconceptualized as "voting with your dollar" as they have been under neoliberal governance? There is mounting evidence that the ethics practised through consumption result in less than favourable outcomes. In fact, many social scientists have found that promoting ethics through shopping results in encouraging the cultural ideology of consumerism, denying the political-economic inequality between social classes, and leaving many pressing environmental issues unaddressed.[11] There are thus good reasons to doubt whether individual market choices can result in significant social and environmental change.

For this reason, it is concerning that scholars have noted a shift under neoliberal governance whereby people are no longer addressed as citizens, but rather are understood, and are being prompted to understand themselves, first and foremost as consumers. Rachel Slocum in the U.S. and John Clarke and Janet Newman in the UK have documented this shift in people's engagements with their local communities and state services, respectively.[12] Indeed, as Rose emphatically shows, in advanced liberalism, consumerism and the logic of choice extend themselves to *all* aspects of social behaviour so that people are expected to use calculating economic behaviour, previously reserved for the marketplace, in all interactions everywhere.[13]

What scholars of governmentality such as Rose and Dean contribute to the literature on consumer choice is careful attention to the ways that agency is reworked in advanced liberalism for the consuming subject.[14] With clear ties to the economic subject of interests (homo economicus) of nineteenth-century liberalism, the neoliberal subject is an entrepreneur of herself. The interests that she registers on non-discriminating markets (as in rational choice models) are now expected to change based on her capacity to be influenced by her environment. She is continually engaged in acquiring new skills and making active choices that will influence all aspects of her future—psychic, material, social, etc. Calculating actions, weighing costs and benefits, investing in the future, and accounting for external contingencies characterize the neoliberal subject active in governing herself through the mechanism of choice. This is a subjectivity that draws on the assumptions of liberal subjectivity, but that intensifies expectations of flexibility and change.

ROUNDUP READY WHEAT AS A MATTER OF INDIVIDUAL MARKET CHOICE

The idea that "voting with your dollar" in the market is the only appropriate site and mechanism for social change was abundantly clear in my interviews with proponents of RR wheat. Such supporters argued that any collective or political decisions about RR wheat would unfairly impede individuals from making their own market decisions, whatever the shortcomings of the crop. In fact, in interviews with me, RR wheat proponents did not even attempt to refute the widely publicized and diverse criticisms that surround the debate about genetic modification. In a move that might mark a shift away from earlier public-relations campaigns where detractors were painted as anti-progress and products of genetic modification were defended as environmentally beneficial,[15] proponents of RR wheat largely accepted that anti-GM movements had serious complaints about RR wheat. Few felt it necessary to convince me that the health and environmental risks associated with GMOs are overblown, or that the corporate control associated with GM crops is benign. On the contrary, even the most vocal supporters of RR wheat, from the Canola Council of Canada and the Canadian Food Inspection Agency, admitted that seed companies pursue their own interests and do not produce the type of traits that farmers find most useful. The widespread criticisms of GM crops were, for proponents, beside the point. Instead, they argued that it is up to individuals to weigh their concerns against any possible private benefits; the risks associated with RR wheat should not preclude private assessments of its merits.

Participants representing the Canadian Food Inspection Agency, Croplife Canada, the Western Canadian Wheat Growers, the Western Barley Growers Association Agriculture and Agri-Food Canada, the Grain Growers of Canada, Monsanto, the Canadian Biotechnology

Advisory Committee, the Canola Council of Canada, Saskatchewan Agriculture and Food, and Agwest Bio argued that the fairest method through which to decide the fate of RR wheat would have been for individual farmers and consumers to vote with their dollars in the marketplace. For example, in an interview with me a representative of the Grain Growers of Canada juxtaposed the voluntary and impartial approach attained through the mechanism of the market with government decision-making processes that were perceived as ineffective and biased. According to the interviewee, the Grain Growers of Canada argued against government involvement (including stakeholder consultations) in the debate about the future of RR wheat. Instead, they hoped for (and ultimately got) "a voluntary option...an industry-driven approach."

Rather than a matter for public policy, the adoption of RR wheat was understood by its proponents as an individual business decision to be left to those whose families and economic well-being depended on the profitability of their farms. For a participant representing the Grain Growers of Canada the decision was "just a business decision—no more, no less, and that's it." Furthermore, they felt it was a decision that legitimately belonged with the farm owner and affected the future of the farm family. Similarly, a representative of the Western Canadian Wheat Growers underscored his organization's approach to the issue as "working with those companies and seeing it as having solutions that producers may choose to utilize or not, and that was a choice for farmers to make." For RR wheat proponents the farmer is the privileged actor and is best able to make decisions at the scale of the individual family.

The logic that markets rather than political movements and governments should decide the fate of RR wheat was also applied to food consumers. A representative from the CFIA explained that as long as any product passes the agency's health and environmental safety risk analysis it can be sold in Canada,[16] and wary food consumers would have to register their concerns through the market. This same respondent elaborated that no consumer is being forced to buy GMOs and that it is always a consumer's right not to buy what s/he does not want to eat. A member of the Canadian Biotechnology Advisory Committee provided further rationale for the consumer friendliness of the market mechanism by arguing that all the right incentives exist for a food company to regulate its own consumer safety. This interviewee argued that because of liability laws, there is no incentive for a company to produce dangerous or ineffective products, since the company could be sued and/or its product could fail in the market. As they told me, "They're not just betting the product line, they're betting the company and so in fact they usually exceed the requirements of the regulatory regime, at least the big ones, because they know there's no tolerance for failure." For supporters of GM technologies, precluding the introduction of RR wheat onto the market unfairly punished companies and consumers who wanted to take advantage of the possible individual rewards associated with genetic modification.

Closely related to the notion of markets as the only reasonable arbiters of individual choice was the conviction that market dynamics support progress. Citing examples of technological development, proponents of RR wheat argued against the use of market impact assessment by regulators in evaluating RR wheat. In these market enthusiasts' views, preventing the introduction of RR wheat would threaten progress in the crop and would be much more harmful to the wheat industry than the loss of export wheat markets. According to this logic, producers should simply adjust to, rather than resist, changing market conditions. A participant from the Canadian Biotechnology Advisory Committee said in our interview: "We don't compensate people who are losing... When CDs, or DVDs became the standard, we didn't compensate the Betamax people who couldn't get videos anymore, we didn't compensate the movie theatres for the fact that they couldn't sell seats anymore. Those resources had to be reallocated...It's just accelerated depreciation and then it's a wash. So you didn't get as much benefit out of it as you might have, that's just the way the world works. That's the mentality that we now have, and it has real power, *because it pushes things forward*" (emphasis added). The debate over market impact was brought to the fore in 2001 when AAFC scientists uncovered a clause in CFIA's regulations that would allow Variety Recommending Committees to reject a new variety based on risks to existing markets.[17] Charged with the task of regulating plants with novel traits, the CFIA had no prior experience evaluating economic risks, and has always insisted that it regulates based on "sound science," narrowly confined to biological impacts. Once this clause was discovered and publicized, farm groups immediately began lobbying CFIA Recommending Committees (that include producer participation and representation), asking them to use this clause in order to reject the introduction of RR wheat if it were to arrive at their committee. Despite considerable support in the farm community for taking economic risks into consideration as part of the evaluation process for new varieties, the CFIA removed the clause in 2002. According to my informant at the CFIA, economic impacts were perceived as beyond the mandate of regulators, who were to ensure the food and environmental safety of a new variety and nothing else. In effect, this decision reinforced the idea that markets, rather than public policy, could best deal with the competing claims and interests among producers and between producers and industry.

AGAINST THE MARKET

The discourses of markets as just mechanisms for registering and ensuring the right of individual choice, as already encompassing the correct incentives to ensure food and environmental safety, and as motivating technological progress, proved hollow at best. Opponents of RR wheat worked hard to convey this to the public by pointing to the limited options that were available on the market, the potential and real harm of RR crops to already existing production systems, and the fact that GM foods are not identifiable in the marketplace. Indeed, proponents of RR wheat themselves often undermined their own arguments for free and just markets in their interviews with me. An excerpt from an interview with a representative from Croplife Canada (a trade association representing numerous plant biotech companies) illustrates this well:

> Right now canola is moving...from open-pollinated, which is where farmers can save their seed...to hybrid seed because they get better yields and better return on their investment...The choice is there...a farmer can choose to grow an open-pollinated variety, but increasingly hybrids are what the farmers are buying because they get better yields, they've got better traits, because again the research and the development is going into the hybrids where the company can capture its investment. So, just like you and I buying quality products or CDs or anything like that, if the artist doesn't get the money back from what they've produced then they can't produce anymore.

Here the interviewee uses the example of hybrid varieties, instead of GM varieties, in order to make the point that farmers, through their market actions, are determining which varieties succeed and fail. In the initial section of the quotation this participant presents the planting of hybrid versus open-pollinated seed as the individual choice of the farmer. However, in the next breath the participant goes on to explain how the existence of hybrid crops is directly dependent on the concentration of resources and research for their development based on their potential to earn corporate profit. While farmers may have the opportunity to choose between the products on the market, their spectrum of choices is narrowly constrained by the capacity of potential varieties to earn profits for agricultural corporations. In a breeding environment where patents on genes are increasingly the norm, products of genetic modification promise opportunities for enhanced accumulation of corporate profit. Thus, like hybrids, GM crops are overrepresented in the spectrum of market choices.

Another proponent of genetic modification, representing the Canola Council of Canada, was equally contradictory in his enthusiasm for market choice, celebrating that "in canola there are really four systems available...so farmers absolutely have a choice about what chemistries they want to and don't want to use on their farm." The respondent here asserts the importance of individual farmers choosing the production systems that best suit them and claims that plenty of agronomic options exist, adding, "If they are relying very heavily on one product... they're probably doing that because it's the most cost-effective option for them." However, the choice that he celebrates involves only four options, controlled and marketed as complete management systems by four large agro-chemical companies. In fact, a farmer representing the same organization, who also supported the introduction of RR wheat, pointed out that the packages promoted by such companies actually preclude agronomic options that might be helpful to farmers and that are normally funded by public research. This interviewee characterized commercial research as only exploiting "short term rewards," whereas government-funded and farmer-funded research is more likely to tackle long-term issues such as disease issues and protein levels in wheat. The farmer explained that the "value" that is housed in output traits such as disease resistance and protein levels is "captured" by the producer, whereas the value associated with herbicide tolerance is captured by the commercial company. Commercial companies produce traits that are part of "agronomic packages" rather than housed in the seed itself. The interviewee continued: "And of course farmers are far more interested in the output traits in the seed because you don't have to then drive to the local farm supply centre to buy $10,000 worth of chemicals."

In direct opposition to the industry's insistence on market choice, farm organizations characterized this narrow set of GM seed varieties, revolving almost singularly around herbicide resistance and marketed by just a few large companies, as a lack of choice. In fact, a representative of the National Farmers Union emphasized to me that at the same time as yields are improving because of the hybridization of canola the spectrum of seed choice is dwindling. After all, farmers cannot buy seeds that retailers are not offering. For this NFU representative the result of narrowed choice is the deskilling of the farmer and a loss of knowledge about biodiversity and productive practices. In this sense, the lack of choice that exists currently in the market shapes future spectrums of choice. Once farmers have lost the ability and tradition of saving seed and selecting for characteristics that suit their local environments, future options are narrowed. Rather than a multiplicity of traits for which farmers select, herbicide resistance comes to dominate the market and professional breeding agendas.

The second argument that opponents of RR wheat used to counter the discourse of individual market choice emphasized the threats that RR crops pose to existing systems of production. The argument that RR wheat threatened current agriculture, and thus narrowed the spectrum of choice for farmers, was advanced through at least two examples. First, farmers and the Canadian Wheat Board felt that the introduction of RR wheat threatened existing wheat markets, especially in Japan and the EU. In fact, the general public was quite sympathetic to the efforts of AAFC and farmers to invoke market impact assessment during the variety recommending process. This short excerpt from an op-ed piece published in the *Globe and Mail* by four professors of agricultural economics, applied microbiology, and food science summarizes the argument nicely:

It would seem logical to adopt a strategy of letting wheat farmers choose between growing GM and non-GM wheat, depending on market signals. For one thing, GM wheat will provide agronomic benefits to some wheat producers. As for the price of GM wheat—which we initially would expect to be lower than non-GM because of consumer resistance—the market will sort out how much of each type is produced to best satisfy its requirements. The trouble with this strategy is that it depends on farmers' ability to segregate the two types of wheat. But farmers' experience with GM canola shows how tricky that can be. And there's virtual consensus in the scientific community that it would be costly and difficult to keep GM and non-GM wheat separate for long.[18]

The second example of RR wheat threatening existing systems of production involved the possibility of maintaining wheat as an organic crop. Opponents of RR wheat, and especially those supporting and involved in organic production, pointed to the loss of canola as an organic crop through widespread contamination of seed stocks as evidence of the non-compatibility of RR and organic systems. On the other hand, I interviewed proponents from the Western Barley Growers Association and the Canadian Food Inspection Agency who maintained that organic producers should bear the responsibility for keeping genetically modified material out of their crops since they were reaping the price premiums of a niche market and self-imposing more strict production standards. Here again, proponents were mobilizing the rhetoric of individual market choice to argue that organic farming was a choice for which farmers accept individual responsibility in meeting the associated standards.

Farm organizations fought back. The Saskatchewan Organic Directorate (SOD), for example, attempted to launch a class action lawsuit seeking compensation for the loss of canola as an organic crop and an injunction against the introduction of GM wheat. In their view, the capacity to produce organically was not an individual choice, but rather the product of a longstanding movement under threat. An interviewee from the SOD said,

> Organic farmers....had the ability and the tradition of being able to supply non-GMO crops and food to the public...go[ing] back thousands of years to the dawn of agriculture. And when an upstart like the biotech companies come along and destroy that ability they should be held accountable. I think the Canadian public should be outraged that their ability to choose to eat non-GMO food is being destroyed...we've had this tradition, this history, this ability to eat non-GMO food, and it's like saying an oil company had a spill in this river, but you know the damage has been done, and so we're just going to sit back and let a certain amount of damage happen every year because...it's just part of modern life, so suck it up.

The irreversibility invoked in this last quotation is indeed very real for farm organizations like the SOD. In fact, according to a representative from the SOD, Rene van Acker, an expert in plant agriculture at the University of Guelph, advised the SOD that contamination of canola with GM material is, all things considered, permanent. In order to "decontaminate" canola, all farms would need to be banned from growing the crop for a period of time. Even then it could end up being "re-contaminated" by residual plants growing in ditches, bushes, or elsewhere. Drawing on the expertise of scientists and the accumulated knowledge associated with widespread contamination of canola on the prairies, anti-RR wheat activists argued that the introduction of RR wheat would threaten the entire wheat industry and all wheat farmers.

The argument that gained the most traction against the champion-ing of market choice revolved around the deliberate withholding of con-sumer labelling of GM products. Most importantly, for the market to perform as a just mechanism for individual consumer choice, consumers of food must have access to full knowledge. Thus, anti-RR wheat activ-ists pointed time and again to Canada's rejection of mandatory labelling legislation for GMOs in order to support their claims that the fate of RR wheat should not be decided in the market. For example, a representa-tive from the Council of Canadians emphasized that it was very easy for the organization to mobilize food consumers and "the average citizen" on the issue since consumers had not asked for the product, there was no perceived need for it, and consumers felt that food was being used as a vehicle for Monsanto's chemicals without the consumer's knowledge.

Indeed, mandatory labelling of GM products had widespread public support. For example, in 2003 the Consumers' Association of Canada made public the results of a national poll that found that 91 percent of Canadian consumers wanted government-enforced labelling on all GM products.[19] This result flew in the face of the voluntary labelling stan-dards upon which the Canadian General Standards Board had finally agreed just months before. The establishment of the voluntary standard was mired in controversy and took a full four years to negotiate. For example, consumer advocates such as the Consumers' Association of Canada walked away from discussions because the possibility of manda-tory labelling had been foreclosed from the beginning.

At the same time as the negotiations for voluntary labelling were taking place, Liberal Member of Parliament (MP) Charles Caccia in-troduced a private member's bill requiring mandatory labelling in 2002. The bill gained considerable momentum in the House of Commons and looked like it might have gone to, and passed, a parliamentary vote under a newly agreed-upon proposal that would have sent all private member's bills worthy of House debate to a vote. Instead, a committee of MPs decided the bill would be debated, but not voted on, in December, just a month before the new voting practice was to be implemented. The possibility of mandatory labelling was thwarted once again, and this seemed to confirm activist claims that the government was beholden to the biotech industry.

According to anti-RR wheat activists, the market mechanism could not be understood as a just arbiter of consumer preferences, especially when the information that food consumers needed to register their opposition to GMOs was withheld. Two strategic moves that quelled any possibility for resistance were in plain view. First, the mechanism for consumer agency was placed in the market rather than in the realms of public policy and social movements. Second, the possibility of opposing GMOs, even through market action, was squashed through the deliberate non-identification of GM products. Needless to say, voluntary labelling has "failed to catch on" among food companies and retailers in Canada.[20]

POLITICAL SUBJECTIVITIES

In order to counter the discourses of consumer choice advanced by proponents of RR wheat, anti-RR wheat activists highlighted their history of collective action in their interviews with me. Participants used two main examples of their capacity to act collectively. First, they spoke of organizing around wheat in the early 1900s when many of the cooperative institutions associated with wheat were formed. Second, participants emphasized the case of coordinated industry action that resulted in the retraction of a genetically modified flax at the turn of the twenty-first century as a counterpoint to the "let the markets decide" approach to GM wheat. It should be taken into account that the form of populist collective action that characterized early-twentieth-century farm organizing was limited, primarily, to the realms of marketing and credit. Farmers remained vehemently protective of private property in land and the exploitation of their own (individual and familial) labour.

Practices of collectivity were accepted and embraced only to the extent that they did not undermine the independent landowning farmer—the master of his own domain. Neither example of collectivity that I mobilize in this section is radical, but both still do the work of denaturalizing the notion of agency as necessarily individual.

It is not a surprise that some of my participants cast back to the first half of the twentieth century when many of the cooperative institutions associated with wheat were formed. In fact, talk of past rounds of collective action in agriculture usually surfaced during interviews with farmers when I asked questions about their involvement in farm politics and when I enquired into the lack of resistance around GM canola. A few participants emphasized that their families had roots in the cooperative movement or with the National Farmers Union and that they brought this inherited experience and understanding to organizing against RR wheat. The following participant from the National Farmers Union thought it imperative to communicate that prairie farm history is a history of collective struggle and cooperation. Any attempt by contemporary farmers to understand themselves as individual entrepreneurial subjects is only possible by erasing the past and the institutional legacy of collective action: "[There has been a] shift from farmers seeing themselves as having a collective interest into one where they really adopted a mythology about how they came and developed here as sort of entrepreneurs on the frontier. Rather [they really have had] a lot of institutional things in place, and a requirement for cooperatives, and a requirement for governments, and a requirement for things like the Manitoba Grains Act, and the weight of the Canadian Seeds Act, and the whole construction to allow them to prosper, and the Canadian Wheat Board being one of them." This quotation begins with an explicit reference to farmers as having collective interests. Here it is not just that producers have the same interests; rather, their welfare is explicitly intertwined through common structures (seed acts, marketing boards, etc.) and experiences.

Such common interests are not the aggregation of interests and preferences at the individual level, as is assumed in the models of public and rational choice reviewed above. They are the result of interconnected practices where the conduct of one or many farmers affects the conduct of others. For example, to the extent that a group of farmers sells their commodities below the market price, the bargaining capacity of all producers of those commodities is affected.

The above participant describes producers as being able to act collectively in order to build and secure institutional supports for their common good. Here agency can be understood as the *product* of relationships and intersubjectivity. Subjects do not come to the public arena with fully formed preferences that can be sufficiently fulfilled through the market mechanism; rather, their preferences are forged in and through their social lives. Production is thus understood as fundamentally social, even if, as the participant describes above, a mythology exists about farmers as individual entrepreneurs tackling the frontier in isolation. Certainly, the spatial arrangement of production in the early-twentieth century (with individual family units producing on separate homesteads) imposed certain barriers to collectivity. But producers did labour together; they often shared equipment and worked each others' land in teams, and they built and relied on cooperative marketing, distribution, and credit structures.

A more recent and topically relevant example of collective action involved coordinated industry retraction of GM flax in the early 2000s. Interestingly, the industry's rejection of the herbicide-residue resistant GM flax (named Triffid) developed at the Crop Development Centre at the University of Saskatchewan received very little press or social-movement activity. Instead, deregistration occurred at the request of the flax industry itself through the influence of the Saskatchewan Flax Development Commission, the Flax Council of Canada, and several farm groups. On one hand, some of the story of GM flax mirrors that of RR wheat. Most importantly for both cases (and for GM crops more broadly),[21] the widespread propagation of GM crops threatened export markets—especially those in Europe, which constituted roughly 60 percent of Canadian flax exports.[22] When European buyers announced, in the summer of 2000, that they would not be buying GM flax, farmers worked through their industry groups to come to the decision that the whole industry should abandon Triffid flax. Moreover, an interviewee from the NFU explained that the particular variety of modified flax did not provide great agronomic benefits to growers since it was resistant to a herbicide that was not highly used and did not offer an advantage in terms of yield. According to this participant, the pushback against GM flax had to be done "on the purely economic market argument" because of the "rightist philosophical inertia" that caused farmers and the industry to resist the pushback.

On the other hand, the story of GM flax played out very differently from that of RR wheat. Unlike RR wheat, GM flax had already successfully emerged from the regulatory process (it was given approval by the CFIA in 1996) and was being reproduced for commercial sale by seed growers all over the prairies when it was deregistered. In fact, 200,000 bushels of seed worth $2.5 million had to be rounded up and crushed in order to destroy the possibility of the seed reproducing and contaminating the environment.[23] Furthermore, farm and industry organizations were opposing a crop that was developed by a public institution, rather than a private company like Monsanto. Indeed, once the decision had been made among the producer organizations that Triffid flax had to be abandoned, the industry was able to put pressure on the developer to voluntarily deregister the variety. Given that the Crop Development Centre received producer check-offs for flax research from the Flax Development Commission, and understood itself as serving farmers, the Centre complied with the industry's wishes. Participants felt that the Crop Development Centre acted reasonably responsibly with regard to the industry's non-acceptance. Despite the fact that both the Crop Development Centre and the seed company had already invested a lot of money in developing and commercializing GM flax, they agreed to retract the product. According to one respondent from the Canola Council of Canada, "while they weren't initially excited about doing it....at the end of the day they did what was best for the industry."

The participants quoted above use the case of coordinated industry action to work against the logic of market choice. This is most obvious in the interview from the representative of the NFU since he frames his discussion of opposition to GM flax as the practice of resistance against the "philosophy" that says "the markets should decide all." Although this participant emphasizes the constrained discursive terrain upon which an articulated resistance had to be constructed (it had to be done "on the purely economic argument"), he underscores a negotiated and intersubjective notion of collectivity. Similarly, the participant from the CCC identifies a coherent unity in "the industry," but this unity is comprised of differently positioned and interested actors, including seed companies and farmers. The industry is clearly not the aggregation of individual fully formed interests and preferences. Instead, it is a site of struggle and contention that is always in the making. In order for GM flax to become deregistered, farm groups had to make a case and represent their arguments in relationship to the economic interests of the industry. In other words, they had to agree and act on a "common good."

The practice of political subjectivity in both examples of collective action described above is social: it is oriented around the possibility of action that supports a negotiated, yet fraught, common good. While this common good may not necessarily represent a broader public good, it is still the result of collective action and a conscious decision made in relation to, but not through, the aggregating mechanism of the market. This is radically different from the political subjectivity associated with consumer choice that posits agency as an individual calculation of costs and benefits.

Consumer choice supports a notion of subjectivity that is fundamentally asocial in the sense that what is best for the sum of individuals is best for society: there is no need for a public sphere, for negotiation, or for a conception of a common good. By advocating that the appropriate mechanism for registering opposition to RR wheat is to vote with your buck in the market, RR wheat proponents effectively denied the commonality of farmers as producers of food. Importantly, when agency is exercised by voting with your dollar on the market, only consumers (either as consumers of food or consumers of farm inputs) can have their say. There exists no possibility for a politics of production since farmers can only register their dissent as consumers of farm inputs. While the farmer understands him/herself first and foremost as a producer, entering the market in order to buy the necessary factors of production, seed and fertilizer companies understand him/her chiefly as a consumer. In *Grundrisse*, Karl Marx outlines exactly this process with regard to the industrial labourer vis-à-vis the capitalist: "What precisely distinguishes capital from the master-servant relation is that the *worker* confronts him as consumer and possessor of exchange values, and that in the form of the *possessor of money*, in the form of money he becomes a simple centre of circulation—one of its infinitely many centres, in which his specificity as worker is extinguished" (emphasis in original).[24] While the relationship of input corporations to farmers is not one of capitalist to labourer, the implications in the two cases are similar. For biotech lobby groups, corporations like Monsanto, and other RR wheat supporters, farmers are simply "centres of circulation." They have agency only inasmuch as they make free choices in the market. Indeed, their specificity as producers/workers is extinguished.

Given Marx's insightful observations in the nineteenth century there seems to be nothing new about the discourse of individual market choice advocated by proponents of RR wheat. However, those studying neoliberalism have shown that the imperative of market choice increasingly pervades more aspects of social life and has become central to the broader concept of freedom. That RR wheat supporters adopted the discourse of farmers as consumers (a very specific positionality of farmers in relation to input suppliers) reflects the incursion of market choice into more and more aspects of social and political life. The capacity to choose through market action was represented as the practice of freedom itself. In this line of thinking individuals understand themselves as entrepreneurs, obliged to navigate through, and demand, an ongoing series of choices that make them who they are. As the quotation from the NFU representative above demonstrates, even past forms of commonality get reconceptualized through this lens. The subjectivity of farmers as collective actors and producers is extinguished in both the past and the present.

Neoliberal subjectivity may be widespread, but it is not total. The two examples of collective action presented above highlight that producers have a wide range of experiences to help them think beyond neoliberal subjectivity. They have both historical and contemporary examples ranging from more radical to quite mainstream, including the type of coordinated industry action that forced the deregistration of GM flax. These memories of working toward a common good inspired farmers in their resistance to RR wheat and allowed them to debunk the idea that individual market choice is the only way of exercising freedom.

Looking back at the history of consumer activism, it is clear that non-market, collective forms of agency do exist for food consumers as well. The notion of "voting with your dollar" that currently dominates Western conceptions of consumer agency also strips consumers of a notion of commonality or collectivity. But consumers can and have worked collectively through actions such as boycotts and protests, through consumer cooperatives, and as allies of producers. Remembering that collective forms of action have existed and do exist for both producers and consumers might enable more creative ways of thinking beyond neoliberal subjectivity in the future.

What's next
FOR ANTI-GM WHEAT POLITICS?

Canadian and American farmers, working in coalition with environmental and health organizations, successfully thwarted Monsanto's first attempts to introduce Roundup Ready wheat. This was a major victory for the anti-GM movement, and it is a fairly stark contrast to the plethora of other genetically modified organisms that have been approved for unconfined release in both Canada and the U.S. over the past decade. However, the story of GM wheat is not over. On 14 May 2009 nine industry groups from Canada, Australia, and the U.S. issued a joint statement calling for the synchronized commercialization of biotech traits in wheat. According to these organizations wheat was suffering from a lack of innovation due to deficient public and private investment. The three signatories from each country (including the Grain Growers of Canada, the Western Canadian Wheat Growers Association, and the Alberta Winter Wheat Producers Commission) argued that low levels of investment had resulted in lower productivity and higher input use in wheat compared to other biotech crops. The statement reflected the interests of industry organizations rather than farmers. In particular, Australia's participation was unsurprising, given that the biotech industry shifted its investments in wheat away from the U.S. and Canada in favour of Australia in the wake of Monsanto's unsuccessful attempt to commercialize RR wheat in North America. In order for Australian GM wheat to be accepted in international markets, there would have to be no easy way for buyers to switch to large GM-free wheat sources. If Canada, the U.S., and Australia all jointly and simultaneously accepted GM wheat there would be more pressure on importing countries to accommodate GM wheat or at least tolerate some level of GM contamination in all wheat.

The participation of three Canadian organizations in the call for synchronized commercialization of biotech wheat demonstrated no shift in public or producer opinion, since these signatories supported the introduction of RR wheat back in the early 2000s and are not considered broad-based farm organizations. The U.S. Wheat Associates (USWA), the North American Millers Association (NAMA), and the National Association of Wheat Growers (NAWG) do appear to have changed their tune on GM wheat, since they all participated in pressuring Monsanto to abandon RR wheat. It is important to note that the USWA and the NAMA are not farm-based organizations; instead they represent international marketing and industrial milling interests. The NAWG is an association of affiliated state-level producer organizations that has taken a decidedly neoliberal turn, which included a change in their vision and mission statements in 2008. Nevertheless, that these organizations are now actively advocating for GM wheat is further evidence that we can expect to find different politics based on particular biotech traits. In fact, in the statement released by the nine organizations there is no mention of Roundup resistance. Instead, these groups suggest that traits such as frost and drought resistance, and enhanced use of soil nutrients and water, might be ripe for biotech development.

In response to the call for synchronized commercialization of biotech wheat, Canadian, American, and Australian farm, environmental, and consumer organizations released a statement in June 2009 reiterating their "definitive rejection of genetically engineered wheat."

Organizations from around the world were asked to sign on to the statement, resulting in 233 endorsements from groups in twenty-six countries. The statement is a categorical rejection of GM wheat that frames genetic engineering as a means for multinational seed companies to strip farmers of their capacity to reproduce seed outside of the market. Specifically, the statement rejects the idea that multinational seed companies are interested in developing traits of basic agronomic importance to farmers; this has been left to public and farmer-funded research and breeding. Furthermore, biotechnology is posited as contrary to the goals of food sovereignty, and thus bad for consumers and civil society more broadly. While a few of the key organizations that led the Canadian coalition against RR wheat are signatories of the counter-statement (including the NFU, the SOD, and Greenpeace), conspicuously absent are general farm organizations like APAS and the Keystone Agricultural Producers. Once again, this shows that broad-based farm organizations are not ready to oppose all forms of genetic modification. However, the presence of non-mainstream farm organizations from all three countries, including the Network of Concerned Farmers (Australia), the Organic Federation of Australia, Biological Farmers of Australia, and the National Family Farm Coalition (USA) demonstrates a level of producer opposition to GM wheat that the nine organizations' call for synchronized commercialization of GM wheat cannot negate.

Meanwhile, Bayer Crop Science and Australia's Commonwealth Scientific and Industrial Research Organisation are busily at work engineering wheat varieties with resistance to drought, more efficient use of soil nutrients, and higher yields. China is also developing GM wheat varieties. In 2009 Monsanto acquired WestBred, LLC, a U.S.-based company that specializes in wheat germplasm. BASF Plant Science and Syngenta have also made recent investments in GM wheat traits. GM wheat varieties in the pipeline are expected to reach commercial regulatory processes around 2020 in North America and Australia. According to my informant from Monsanto, the company is not likely to abandon the RR trait, even though it is explicitly avoiding talking about RR. Instead, it is likely that Monsanto will stack other traits on top of the RR trait in order to make the GM package more appealing to customers (both consumers and producers), while locking farmers into buying Monsanto seed and herbicide.

The politics of opposition to GM wheat crosses some complicated terrain that makes it difficult to reconcile categorical fears about Frankenfoods with questions about the control and commodification of seed or the very local and specific needs of different crops in farmers' fields. The strategic focuses of the anti-RR wheat coalition on markets and the particular RR trait might not sustain another round of opposition to new GM wheat traits. For example, biotech companies may introduce "second generation" traits that are more appealing to the consuming public, including "functional foods" enriched with vitamins or cancer-fighting agents. Or they may introduce modifications that would be more appealing to farmers, including drought resistance. Additionally, work is currently underway on finding suitable segregation systems that could keep GM and non-GM varieties separated throughout supply chains and thus allow the two conventions to "coexist." If Europe and Japan decide to tolerate some low level of GM contamination, segregation will become an increasingly viable option. In other words, the very local and specific concerns with particular crops and traits provides an uncertain future for farmers' engagement with anti-GM politics.

In speculating about the future it is easy to lose sight of a fact that often goes unchallenged: the increasing corporate control over agricultural systems. This fact was brought into sharp focus in 2009 when GM Triffid flax, pulled from Canadian fields and destroyed in 2001, appeared in Canadian exports to Europe. Recall that Canadian flax growers were the driving force behind the deregistration of Triffid before it had been widely grown across the prairies. Testing conducted in 2009 revealed widespread contamination of flax with GM material at low levels, with no geographical concentration. Authorities are still unsure about exactly how such widespread contamination happened. Given closed export markets in Europe, Sri Lanka, Korea, Thailand, and Japan for the 2010 growing season, grain companies, including Viterra, tried to force Canadian flax growers to abandon their farm-saved seed, claiming that only certified commercially available seed could be guaranteed GM-free. According to the Canadian Biotechnology Action Network approximately 75 percent of Canada's flax farmers use farm-saved seed. The Flax Council of Canada (FCC) initially agreed that all flax should be grown from certified seed. But Canadian flax farmers fought back once again and demanded that they be able to use farm-saved seed. In response, the FCC established a domestic stewardship program that included an option for producers to use farm-saved seed provided it underwent the same rigorous sampling and testing procedures as commercially available certified seed.

The resurfacing of Triffid flax eight years after it was systematically rounded up and destroyed from seed growers is a cautionary tale for GM wheat. First, it demonstrates the absurdity of keeping GM and non-GM crops separate. Even if separate handling systems could be established for GM wheat there is no plausible way of containing seeds grown in open fields and transported by leaky trucks and railcars. But the plausibility of viable segregation is beside the point. Once widespread contamination occurs there is no putting the genie back in the bottle, and markets around the world will have to accept contamination. The biotech industry will then have carte blanche to market its products. Second, the Triffid flax story demonstrates how desperately grain companies want to pry away the capacity of farmers to reproduce seed outside of the market.

If the family farm is going to persist, many organizations involved in the coalition against RR wheat argued that wheat will need to continue to be produced at high standards of quality for human consumption, in viable rotations with other crops, and with some level of producer control over agronomic practice, including the capacity to save and reproduce seed. Wheat, therefore, occupies a prominent position in the prairie agrarian question, such that the corporate takeover of wheat would threaten the capacity of family farms to continue subsisting in ways that have not been the case with the corporate control of other crops. One of the reasons that small farmers have been able to persist as long as they have is that they have built collective marketing and producer organizations that work in their interests. Such institutions were built around mass production and international marketing; any challenge to these mechanisms also threatens to undermine producer organizations. The recent dismantling of the Canadian Wheat Board will deal another blow to the capacity for prairie farmers to organize against genetic modification.

It seems a new round of resistance to GM wheat is urgently required, but the strategies and tactics that successfully warded off GM wheat in the early 2000s will not work again. Anti-GM farm activists will have to deal with international support from some producer organizations for the introduction of GM wheat and re-orient themselves to a post-single desk environment. One of their strongest coalition members, the CWB, is now significantly weakened and grain handling corporations are rushing in. Given this new environment producers must articulate with more force and clarity that the main problem with GMOs (and indeed the food system more broadly) is its corporate control. By arguing that farmers' and the public's abilities to control our food systems is being aggressively eroded by corporate power, necessarily local and specific struggles can be linked up to a more global resistance. The corporate control of biotechnology is a discourse behind which both producers and consumers can unite, and it names a problem that requires collective action in order for it to be challenged. Such an analysis of the problem with biotechnology also allows for the possibility of coalition work with other people struggling over agriculture and food across the globe. While opposition to individual modifications in particular crops will need to play out through specific campaigns that take into account the contingencies associated with local ecologies, histories, and cultures, a broader and more international struggle is also needed to sustain pressure on multinational companies. In addition, international campaigns against GMOs and corporate control will need to find space in their movements for those producing in the fields who are best positioned to understand local ecological, social, and economic outcomes and to take on specific campaigns in locally relevant ways.

Notes

PREFACE

1 David Harvey, *A Brief History of Neoliberalism* (New York: Oxford University Press, 2005), 2.

2 See for example Jamie Peck and Adam Tickell, "Neoliberalizing Space," *Antipode* 34, 3 (2002): 380–404; Bob Jessop, "Liberalism, Neoliberalism, and Urban Governance: A State-Theoretical Perspective," *Antipode* 34, 3 (2002): 452–71.

3 Nancy Hartsock, "Rethinking Modernism: Minority versus Majority Theories," *Cultural Critique* 7 (1987): 188.

INTRODUCTION

1 "Monsanto to Realign Research Portfolio: Development of Roundup Ready Wheat Deferred," Monsanto Company, news release, 10 May 2004, http://monsanto.mediaroom.com/index.php?s=43&item=241. Roundup is the brand name of a Monsanto herbicide, which is composed mostly of glyphosate. Roundup Ready crops include corn, cotton, soy, and canola. They are genetically engineered to be resistant to Roundup.

2 H.J. Beckie et al., "A Decade of Herbicide-Resistant Crops in Canada," *Canadian Journal of Plant Science* 86, 4 (2006): 1244.

3 See for example M. Fulton et al., *Transforming Agriculture: The Benefits and Costs of Genetically Modified Crops.* (Saskatoon: University of Saskatchewan, 2001).

4   See for example Ann Elizabeth Reisner, "Social Movement Organizations' Re-actions to Genetic Engineering in Agriculture," *American Behavioural Scientist* 44, 8 (2001): 1389–1404; Robin Jane Roff, "Shopping for Change? Neolib-eralizing Activism and the Limits to Eating Non-GMO," *Agriculture and Human Values* 24, 4 (2007): 511–22; Rachel Schurman, "Fighting Frankenfoods: Industry Structures and the Efficacy of the Anti-Biotech Movement in Western Europe," *Social Problems* 51, 2 (2004): 243–68.

5   Biopiracy is the appropriation of the genetic information and biodiversity of the global South by companies from the global North through patenting. See Vandana Shiva, *Biopiracy: The Plunder of Nature and Knowledge* (Cambridge, MA: South End Press, 1997).

CHAPTER ONE: SETTING THE STAGE

1   Tokar, Brian. "A history of biotech giant Monsanto," CBC digital archives, 3 May 1999, http://www.cbc.ca/archives/categories/economy-business/agricul-ture/genetically-modified-food-a-growing-debate/history-of-biotech-giant-monsanto.html.

2   Monsanto Company, "Putting Technology to Work in the Field," 2005, http://www.monsanto.co.uk/news/product_pipeline/pipeline_brochure.pdf.

3   John F. Varty, "On Protein, Prairie Wheat, and Good Bread: Rationalizing Technologies and the Canadian State, 1912–1935," *Canadian Historical Review* 85, 4 (2004): 721–53.

4   Since its settling, populist farm politics in Alberta have taken on a more con-servative undertone than in the other prairie provinces. This is partly the result of diverging political histories and the dominance of the Alberta Social Credit Party starting in the 1930s in the prairie provinces. More generally, Alberta has followeding a less progressive set of policies and stances including vehement opposition to the federal government and widespread belief in small govern-ment more generally. It was no secret that the Alberta government supported the introduction of RR wheat, and this may have had some effect on WRAP's decision to not to join the coalition. Also, wheat farming is of less importance in Alberta than in Manitoba and Saskatchewan (as shown in Table 5), and WRAP advocates that farmers produce less of wheat in favour of value-added products such as boxed beef, processed meat, food products, pharmaceuticals, and biofuels. Finally, although neither the Agricultural Producers Association of Saskatchewan nor the Keystone Agricultural Producers in Manitoba can be considered particularly progressive organizations, a perusal of WRAP's website reveals more support for "research and innovation," a strong denunciation of government involvement in the Canadian Wheat Board, and a more pro-mar-ket approach.

5 The CWB has a long and complex history. Initially established in 1935 as a voluntary government agency for the marketing of wheat, during the Second World War its mandate was extended to the monopoly marketing of all Canadian grains. In 1949 its purview changed again and the CWB became responsible for only wheat, oats, and barley. Oats were removed from CWB jurisdiction in 1989. Importantly, in 1998 the CWB Act was changed to allow "shared governance" between farmers, who elect ten of the fifteen members of the Board of Directors, and the federal government, that appoints the other five (including the president and CEO). The CWB monopoly has faced increasing attack in recent years, especially since the election of the minority Conservative Party government in Canada in early 2006. Most recently, with the election of a majority Conservative Party government in 2011, the CWB was legislated to end its single-desk marketing monopoly in August of 2012. Court challenges to this legislation are still working their way through the legal system.

6 Stephen Ackroyd and John A. Hughes, *Data Collection in Context*, 2nd ed. (London: Longman Group UK, 1992), 103.

7 Robert S. Weiss, *Learning from Strangers: The Art and Method of Qualitative Interview Studies* (New York: The Free Press, 1994), 149.

8 Tim May, *Social Research: Issues, Methods and Process* (Buckingham: Open University Press, 1993), 108.

9 Tim Dant, *Knowledge, Ideology and Discourse: A Sociological Perspective* (London: Routledge, 1991), 235.

10 Ibid., 209.

11 Andrea Fontana and James H. Frey. "The Interview: From Structured Questions to Negotiated Text," in *Handbook of Qualitative Research*, 2nd ed., ed. Norman K. Denzin, and Yvonna. S. Lincoln (Thousand Oaks, CA: Sage, 2000), 645–72.

12 Vladimir Ilyich Lenin, *The Development of Capitalism in Russia* (Moscow: Progress Publishers, 1967).

13 Karl Kautsky, *The Agrarian Question,* trans. Peter Burgess (London: Zwan Publications, 1988).

14 Ibid., 110.

15 Ibid., 324.

16 Alexander V. Chayanov, *The Theory of Peasant Economy*, ed. Daniel Thorner, Basil Kerblay, and R.E.F. Smith (Homewood, IL: Richard D. Irwin, 1966).

17 Ibid., 112.

18 Harriet Friedmann, "Household Production and the National Economy: Concepts for the Analysis of Agrarian Formations," *Journal of Peasant Studies* 7, 2 (1980): 158–84.

19 Susan A. Mann and James M. Dickinson, "Obstacles to the Development of a Capitalist Agriculture," *Journal of Peasant Studies* 5, 4 (1978): 466–81.

20 Susan Mann, *Agrarian Capitalism in Theory and Practice* (Chapel Hill: University of North Carolina Press, 1990).

21 David Goodman, Bernardo Sorj, and John Wilkinson, *From Farming to Biotechnology: A Theory of Agro-Industrial Development* (Oxford: Basil Blackwell, 1987), 7.

22 Ibid., 58.

23 Jack Ralph Kloppenburg, *First the Seed: The Political Economy of Plant Biotechnology,* 2nd ed. (Madison: University of Wisconsin Press, 2004).

24 George L. Henderson, *California and the Fictions of Capital* (New York: Oxford University Press, 1999).

25 Statistics Canada, "Farm Data and Farm Operator Data: 2006 Census of Agriculture," 1 December 2008, http://www.statcan.gc.ca/pub/95-629-x/95-629-x2007000-eng.htm.

CHAPTER 2: REGULATING AND PROMOTING BIOTECHNOLOGY IN CANADA

1 The classification of countries based on precautionary vs. promotional and bottom-up vs. top-down approaches is based on Michael Howlett and Andrea Migone, "Explaining Local Variation in Agri-Food Biotechnology Policies: 'Green' Genomics Regulation in Comparative Perspective," *Science and Public Policy* 37, 10 (2010): 781–95.

2 Elisabeth Abergel and Katherine Barrett, "Putting the Cart before the Horse: A Review of Biotechnology Policy in Canada," *Journal of Canadian Studies* 37, 3 (2002): 135–61.

3 Ibid., 137–38.

4 National Advisory Board on Science and Technology, quoted in Abergel and Barrett, "Putting the Cart," 140–41.

5 Richard Carew, "Science Policy and Agricultural Biotechnology in Canada," *Review of Agricultural Economics* 27, 3 (2005): 300–16.

6 Peter Andree, "The Genetic Engineering Revolution in Agriculture and Food: Strategies of the 'Biotech Bloc,'" in *The Business of Global Environmental Governance*, ed. David L. Levy and Peter J. Newell (Cambridge, MA: MIT Press, 2005), 137.

7 Elizabeth Moore, "The New Direction of Federal Agricultural Research in Canada: From Public Good to Private Gain?," *Journal of Canadian Studies* 37, 3 (2002): 112–34.

8 Sara Bjorkquist and Mark Winfield, *The Regulation of Agricultural Biotechnology in Canada* (Toronto: Canadian Institute for Environmental Law and Policy, 1999), 21, http://cielap.org/pdf/regbiotch.pdf.

9 Abergel and Barrett, "Putting the Cart."

10 Government of Canada, quoted in Abergel and Barrett, "Putting the Cart," 149.

11 Standing Committee on Agriculture and Agri-Food, "rBST in Canada" (Ottawa: Government of Canada, 1994), quoted in Abergel and Barrett, "Putting the Cart," 149.

12 Standing Committee on Environment and Sustainable Development. Biotechnology Regulations in Canada: A Matter of Public Confidence. Ottawa: Government of Canada, 1996 quoted in Abergel and Barrett, "Putting the Cart," 150.

13 National Biotechnology Advisory Committee. Sixth Report. "Leading the Next Millennium" (Ottawa: NBAC, 1998), cited in Abergel and Barrett, "Putting the Cart," 150.

14 Romeo. F. Quijano, "Elements of the Precautionary Principle," in *Precaution, Environmental Science, and Preventive Public Policy*, ed. Joel A. Tickner (Washington: Island Press, 2003), 23–26.

15 Joel A. Tickner, "The Role of Environmental Science in Precautionary Decision Making," in *Precaution, Environmental Science, and Preventive Public Policy*, ed. Joel A. Tickner (Washington: Island Press, 2003), 6.

16 Peter Andrée and Lucy Sharratt, *Genetically Modified Organisms and Precaution: Is the Canadian Government Implementing the Royal Society of Canada's Recommendations?* (Ottawa: The Polaris Institute, 2004).

17 Scott Prudham and Angela Morris, "Regulating the Public to Make the Market 'Safe' for GM Foods: The Case of the Canadian Biotechnology Advisory Committee," *Studies in Political Economy* 78 (2006): 145–75.

18 Ibid., 169.

19 Ibid., 158.

20 Ibid., 163.

21 André Magnan, "Refeudalizing the Public Sphere: 'Manipulated Publicity' in the Canadian Debate on GM Foods," *Canadian Journal of Sociology* 31, 1 (2006): 25–53.

22 Simon Enoch, *The Potemkin Corporation: Corporate Social Responsibility, Public Relations and Crises of Democracy and Ecology* (PhD diss., Ryerson University and York University, 2009).

23 Barry Wilson, "Market Risk Once Part of Process," *Western Producer*, 10 April 2003.

24 Moore, "New Direction."

25 Barry Wilson, "Gov't Limits Debate on CWB Bill," *Western Producer Daily News*, 20 October 2011.

26 In fact the CWB administered a plebiscite in the summer of 2011 that asked whether farmers would like to keep the single-desk marketing agency, with a response rate of 56 percent. Sixty-two percent of respondents voted in favour of keeping the single desk for wheat and 51 percent voted to retain it for barley.

27 Barry Wilson, "Vanclief Floats Nonscientific GM Test," *Western Producer*, 30 October 2003.

28 Canadian Biotehnology Action Network, "Updates: Victory for Democracy in Canada on Genetic Engineering!," 2 December 2010, http://www.cban.ca/Take-Action/Action-Closed-Bill-C-474.

29 Sarah Hartley and Grace Skogstad, "Regulating Genetically Modified Crops and Foods in Canada and the United Kingdom: Democratizing Risk Regulation," *Canadian Public Administration* 48, 3 (2005): 305–27.

30 Barry Wilson, "Labeling Bill Misses Automatic Vote Rule," *Western Producer*, 28 November 2002.

31 Ibid.

32 Hartley and Skogstad, "Regulating."

CHAPTER 3: THE DIFFERENCE BETWEEN BREAD AND OIL

1 Lawrence Busch, William B. Lacy, and Jeffrey Burkhardt, *Plants, Power, and Profit: Social, Economic, and Ethical Consequences of the New Biotechnologies* (Cambridge, MA: Blackwell, 1991), 129.

2 Donna J. Haraway, *Simians, Cyborgs, and Women: The Reinvention of Nature* (New York: Routledge, 1991), 199-201.

3 Donna J. Haraway, *The Haraway Reader* (New York: Routledge, 2004), 328.

4 See for example William Boyd, W. Scott Prudham, and Rachel A. Schurman, "Industrial Dynamics and the Problem of Nature," *Society and Natural Resources* 14, 7 (2001): 555–70.

5　See for example Michael Callon, "Some Elements in a Sociology of Translation: Domestication of the Scallops and Fishermen of St. Brieuc Bay," in *Power, Action, Belief: A New Sociology of Knowledge?* ed. John Law (London: Routledge and Kegan Paul, 1986); Bruno Latour, *We Have Never Been Modern* (Cambridge, MA: Harvard University Press, 1993); Bruno Latour, "When Things Strike Back: A Possible Contribution of 'Science Studies' to the Social Sciences," *British Journal of Sociology* 51, 1 (2000): 107–23; John Law, ed., *A Sociology of Monsters: Essays on Power, Technology and Domination* (London: Routledge, 1991).

6　Timothy Mitchell, *Rule of Experts: Egypt, Techno-Politics, Modernity* (Berkeley: University of California Press, 2002).

7　See for example Jonathan Murdoch, "Inhuman/Nonhuman/Human: Actor-Network Theory and the Prospects for a Nondualistic and Symmetrical Perspective on Nature and Society," *Society and Space* 15, 6 (1997): 731–56; Margaret FitzSimmons and David Goodman, "Incorporating Nature: Environmental Narratives and the Reproduction of Food," in *Remaking Reality: Nature at the Millennium*, ed. Bruce Braun and Noel Castree (London: Routledge, 1998), 194–220; Sarah Whatmore, *Hybrid Geographies: Natures, Cultures, Spaces* (London: Sage Publications, 2002); David Goodman, "Ontology Matters: The Relational Materiality of Nature and Agro-Food Studies," *Sociologia Ruralis* 41, 2 (2001): 182–200.

8　Donna Haraway, "Nature, Politics, and Possibilities: A Debate and Discussion with David Harvey and Donna Haraway," *Environment and Planning D: Society and Space* 13, 5 (1995): 509.

9　Donna Haraway, *When Species Meet* (Minneapolis: University of Minnesota Press, 2008).

10　Haraway, *The Haraway Reader*, 302.

11　George Edwin Britnell, *The Wheat Economy* (Toronto: University of Toronto Press, 1939); Vernon Clifford Fowke, *The National Policy and the Wheat Economy* (Toronto: University of Toronto Press, 1957).

12　Sarah Carter, "Two Acres and a Cow: 'Peasant' Farming for the Indians of the Northwest, 1889–97," *Canadian Historical Review* 70, 1 (1989): 27–52.

13　Homesteads were provided through the Dominion Lands Act of 1872 for a registration fee of ten dollars. In order to keep the quarter section (160 acre) homestead, by the end of three years the settler had to have lived on the land for at least six months of every year, have erected a shelter, and have successfully broken the soil and begun farming crops on 15 acres. John W. Warnock, *Saskatchewan: The Roots of Discontent and Protest* (Montreal: Black Rose Books, 2004). Aboriginal people and any woman who was not the sole head of a family were ineligible to apply. Bill Waiser, *Saskatchewan: A New History* (Calgary: Fifth House, 2005), 104.

14 According to Harriet Friedmann, the world wheat market began to emerge after 1873, spurred on by the export orientation of "new world" agriculture. Interestingly, household production of the kind driving the growth of the wheat economy on the Canadian prairies outcompeted capitalist relations of wheat production which dominated in many countries in Europe well into the 1900s. Harriet Friedmann, "World Market, State, and Family Farm: Social Bases of Household Production in the Era of Wage Labor," *Comparative Studies in Society and History* 20, 4 (1978): 545–86.

15 Fowke, *National Policy*, 72.

16 Britnell, *Wheat Economy,* 33.

17 Fowke, *National Policy*, 71.

18 S. M. Lipset, *Agrarian Socialism: The Cooperative Commonwealth Federation in Saskatchewan, A Study in Political Sociology* (Berkeley: University of California Press, 1950); Dale Eisler. *False Expectations: Politics and the Pursuit of the Saskatchewan Myth* (Regina: Canadian Plains Research Centre, 2006); Warnock, *Saskatchewan.*

19 Friedmann argues that between 1873 and 1935 household production of wheat began to out-compete capitalist production (based on wage labour) that had dominated in wheat-producing regions prior to 1873. This was precisely because households did not *require* a normal rate of profit, and were able to restrict personal consumption and investment in production in ways that capitalist farms, structurally, could not accept. Even in the extreme situation of pioneer prairie farmers, as explained above, households continued to produce without profit. Friedmann, "World Market."

20 Warnock, *Saskatchewan*, 132.

21 Fairbairn, Garry, *From Prairie Roots: The Remarkable Story of Saskatchewan Wheat Pool* (Saskatoon: Western Producer Prairie Book, 1984), 3.

22 Donald Avery, "The Radical Alien and the Winnipeg General Strike of 1919," in *Canadian Working Class History: Selected Readings,* 3rd ed., ed. Laurel Sefton Macdowell and Ian Radforth (Toronto: Canadian Scholars Press, 2006), 217–31.

23 Eisler, *False Expectations*, 62.

24 Lipset, *Agrarian Socialism*, 66.

25 Eisler, *False Expectations*.

26 Fairbairn, "Saskatchewan Co-operative."

27 Paul F. Sharp, *The Agrarian Revolt in Western Canada* (1948; repr., Regina: Canadian Plains Research Centre, University of Regina, 1997), 24.

28 Fairbairn, *From Prairie Roots,* 38.

29 Ibid., 241–42.

30 "About Us: History," Canadian Wheat Board, http://www.cwb.ca/public/en/about/history/ (accessed 12 December 2007).

31 The Conservative Party federal government that came to power in early 2006, initially with a minority mandate, opposed the monopoly character of the CWB and has systematically moved toward the removal of its single desk. In March 2007 it conducted a plebiscite on the marketing of barley, asking Western Canadian farmers whether barley should be removed from the mandate of the CWB. While the government wanted to move forward on removing barley from the mandate of the CWB, controversy erupted about whether it could legally do so without changing the CWB legislation. After being elected with a majority mandate in 2011 the government moved swiftly to eliminate the single desk through legislation that came into effect on 1 August 2012.

32 "About Us," Canadian Wheat Board.

33 Waiser, *Saskatchewan,* 123.

34 As John F. Varty documents, an antagonism exists between protein content and aridity of the growing region such that the spatial expansion enabled by early maturation into less arid regions of the prairies had the effect of reducing protein content in wheat in those regions. High protein content, although not explicitly factored into the grading system, was demanded by flour millers as it made for "better" bread. Varty, "On Protein" (see chap. 1, n. 10).

35 Moore, "New direction" (see chap. 2, n. 40).

36 Britnell, *Wheat Economy,* 48.

37 Breeding programs in wheat have also revolved heavily around increasing yield and on quality characteristics that are of interest to the milling, baking, and processing industries—including protein content, colour, and baking strength.

38 Wheat is understood by plant breeders as a much less promiscuous plant than canola or corn. It typically self-pollinates, although as much as 5 percent outcrossing has been reported in the case of stray pollen. Stephan Symko, *From a Single Seed: Tracing the Marquis Wheat Success Story in Canada to Its Roots in Ukraine* (Ottawa, ON: Research Branch Agriculture and Agri-Food Canada, 1999).

39 Agriculture and Agri-Food Canada, *Profile of the Canadian Wheat Industry,* Bi-Weekly Bulletin 20 (2004): 1–8.

40 Ibid.

41 Eisler, *False Expectations*.

42 Brewster Kneen, *The Rape of Canola* (Toronto: NC Press, 1992), 28.

43 Lawrence Busch and Arunas Juska, "Beyond Political Economy: Actor Networks and the Globalization of Agriculture," *Review of International Political Economy* 4, 4 (1997): 697.

44 Peter W.B. Phillips and George G. Khachatourians, Introduction and Overview, in *The Biotechnology Revolution in Global Agriculture: Invention, Innovation and Investment in the Canola Sector*, ed. Phillips, and Khachatourian (New York: CABI Publishing, 2001), 16–17.

45 Kneen, *Rape of Canola*, 15.

46 George G. Khachatourians, Arthur K. Sumner, and Peter W.B. Phillips, "An Introduction to the History of Canola and the Scientific Basis for Innovation," in Phillips and Khachatourians, *Biotechnology Revolution*, 38.

47 "About Us," Canola Council of Canada, http://www.canola-council.org/council.html (accessed 22 January 2008).

48 M. Fulton et al., *Transforming Agriculture*, 89 (see Introduction, n. 3).

49 As outlined by Kneen, *Rape of Canola*, 18.

50 M. Fulton et al., *Transforming Agriculture*, 92 (see Introduction, n. 3).

51 Ibid., 95.

52 Ibid., 94–95.

53 Kloppenburg, *First the Seed* (see chap. 2, n. 28).

54 M. Fulton et al., *Transforming Agriculture*, 89 (see Introduction, n. 3).

55 Peter W.B. Phillips, "The Role of Private Firms," in Phillips and Khachatourians, *Biotechnology Revolution*, 105–28.

56 Ibid., 119.

57 M. Fulton et al., *Transforming Agriculture* (see Introduction, n. 3).

58 Kloppenburg, *First the Seed* (see chap. 2, n. 28).

59 Canadian Food Inspection Agency, *The Biology of Brassica Napus L.* (Ottawa: Plant Biosafety Office, 1994).

60 Studies conducted by Agriculture and Agri-Food Canada and by University of Manitoba plant scientists found that even pedigreed seed lots for conventional canola had been contaminated with herbicide-resistant traits at levels above 0.25 percent (the threshold above which pedigreed seed cannot be certified). R. Downey and H.J. Beckie, *Isolation Effectiveness in Canola Pedigree Seed Production*. (Saskatoon, SK: Agriculture and Agri-Food Canada, 2002); Lyle F. Fri-

esen, Alison G. Nelson, and Rene C. Van Acker, "Evidence of Contamination of Pedigreed Canola *(Brassica Napus)* Seedlots in Western Canada with Genetically Engineered Herbicide Resistance Traits," *Agronomy Journal* 95, 6 (2003): 1342–47.

61 Phillips, "Private Firms," 107.

CHAPTER 4: FARMERS MAKE THEIR CASE AGAINST GM WHEAT

1   On international regulation see Jennifer Clapp, "Transnational Corporate Interests and Global Environmental Governance: Negotiating Rules for Agricultural Biotechnology and Chemicals," *Environmental Politics* 12, 4 (2003): 1–23; Daniel Lee Kleinman and Abby J. Kinchy, "Against the Neoliberal Steam Roller? The Biosafety Protocol and the Social Regulation of Agricultural Biotechnologies," *Agriculture and Human Values* 24, 2 (2007): 195–206; Makane Moïse Mbengue and Urs P. Thomas, "The Precautionary Principle: Torn between Biodiversity, Environment-Related Food Safety and the WTO," *International Journal of Global Environmental Issues* 5, 1–2 (2005): 36–53; Peter Newell and Ruth Mackenzie, "Whose Rules Rule? Development and the Global Governance of Biotechnology," *IDS Bulletin* 35, 1 (2004): 82–91. On Canadian regulation see Peter Andree, "The Biopolitics of Genetically Modified Organisms in Canada," *Journal of Canadian Studies* 37, 3 (2002): 162–91; Hartley and Skogstad, "Regulating" (see chap. 2, n. 61); Prudham and Morris, "Regulating the Public" (see chap. 2, n. 49); R. Steven Turner, "Of Milk and Mandarins: RBST, Mandated Science and the Canadian Regulatory Style," *Journal of Canadian Studies* 36, 3 (2001): 107–30.

2   On social resistance to GMOs see Clare Hall and Dominic Moran, "Investigating GM Risk Perceptions: A Survey of Anti-GM and Environmental Campaign Group Members," *Journal of Rural Studies* 22, 1 (2006): 29–37; Chaia Heller, "Post-Industrial 'Quality Agricultural Discourse': Techniques of Governance and Resistance in the French Debate over GM crops." *Social Anthropology* 14, 3 (2006): 319–34; Andre Magnan, "Strange Bedfellows: Contentious Coalitions and the Politics of GM Wheat," *The Canadian Review of Sociology and Anthropology* 44, 2 (2007): 289–317; Birgit Müller, "Infringing and Trespassing Plants: Patented Seeds at Dispute in Canada's Courts," *Focaal European Journal of Anthropology* 48, 1 (2006): 83–98; Derrick A. Purdue, *Anti-GenetiX: The Emergence of the Anti-GM Movement* (Aldershot: Ashgate, 2000); Reisner, "Social Movement" (see Introduction, n. 4); Roff, "Shopping" (see Introduction, n. 4); Schurman, "Fighting Frankenfoods" (see Introduction, n. 4); Rachel Schurman and William Munro, "Ideas, Thinkers, and Social Networks: The Process of Grievance Construction in the Anti-Genetic Engineering Movement," *Theo-*

ry and Society 35, 1 (2006): 1–38; Abby Kinchy, *Seeds, Science and Struggle: The Global Politics of Transgenic Crops* (Cambridge, MA: MIT Press, 2012).

3   Schurman, "Fighting Frankenfoods" (see Introduction, n. 4).

4   Schurman and Munro, "Ideas, Thinkers."

5   Hall and Moran, "Investigating GM."

6   Frederick H. Buttel, "The Environmental and Post-Environmental Politics of Genetically Modified Crops and Foods," *Environmental Politics* 14, 3 (2005): 309–23.

7   Purdue, *Anti-GenetiX*, 60.

8   Ibid., 135.

9   Reisner, "Social Movement," 1392–93. (see Introduction, n. 4)

10 Roff, "Shopping." (see Introduction, n. 4)

11 On producer perception of GM crops see, for example, Derek Berwald, Colin A. Carter, and Guillaume P. Gruère, "Rejecting New Technology: The Case of Genetically Modified Wheat," *American Journal of Agricultural Economics* 88, 2 (2006): 432–47; David S. Bullock and Elisavet I. Nitsi, "Roundup Ready Soybean Technology and Farm Production Costs: Measuring the Incentive to Adopt Genetically Modified Seeds," *American Behavioral Scientist* 44, 8 (2001): 1283–1301; Murray Fulton and Lynette Keyowski, "The Producer Benefits of Herbicide-Resistant Canola," *AgBioForum* 2, 1 (1999): 85–93; Peter D. Goldsmith, "Innovation, Supply Chain Control, and the Welfare of Farmers: The Economics of Genetically Modified Seeds," *American Behavioral Scientist* 44, 8 (2001): 1302–26; Clare Hall, "Identifying Farmer Attitudes Towards Genetically Modified (GM) Crops in Scotland: Are They Pro- or Anti-GM?" *Geoforum* 39, 1 (2008): 204–12.

12 Magnan, "Strange Bedfellows"; Müller, "Infringing and Trespassing."

13 Heller, "Post-Industrial."

14 Jonathan Murdoch, Terry Marsden, and Jo Banks, "Quality, Nature, and Embeddedness: Some Theoretical Considerations in the Context of the Food Sector," *Economic Geography* 76, 2 (2000): 107–25.

15 On "quality" production as a characteristic of post-industrial agriculture see for example Terry Marsden and Everard Smith, "Ecological Entrepreneurship: Sustainable Development in Local Communities through Quality Food Production and Local Branding," *Geoforum* 36, 4 (2005): 440–51; Henk Renting, Terry K. Marsden, and Jo Banks, "Understanding Alternative Food Networks: Exploring the Role of Short Food Supply Chains in Rural Development," *Environment and Planning A* 35 (2003): 393–411.

16 Ilbery and Kneafsey outline the turn to post-productivism in Europe: Brian Il-
bery and Moya Kneafsey, "Product and Place: Promoting Quality Products and
Services in the Lagging Regions of the European Union," *European Urban and
Regional Studies* 5, 4 (1998): 329–41.

17 Ibid.

18 Betsy Donald and Alison Blay-Palmer, "The Urban Creative-Food Economy:
Producing Food for the Urban Elite or Social Inclusion Opportunity?" *Environment and Planning A* 38, 10 (2006): 1901–20.

19 Keynesianism was a dominant form of economic organization in the post-war
period in advanced industrialized countries. It is generally acknowledged that
Keynsian economies suffered from widespread crisis in the early 1970s and were
replaced by neoliberal policies during the 1980s. Keynesianism is based on the
ideas of John Maynard Keynes (1883–1946), who advocated for state interven-
tion in the economy in order to smooth out capitalist business cycles of boom
and bust with state spending and a minimum level of welfare for national citizens. Keynesianism sought full employment and understood effective demand
(consumer spending) as capable of driving economic growth. In geography,
there is a general acknowledgement that Keynesian policies also sought equi-
table distribution of economic activity across space.

20 David Goodman and Melanie DuPuis, "Knowing Food and Growing Food:
Beyond the Production-Consumption Debate in the Sociology of Agriculture,"
Sociologia Ruralis 42, 1 (2002): 5–22.

21 Ian Cook and Philip Crang, "The World on a Plate: Culinary Culture, Displacement and Geographical Knowledges," *Journal of Material Culture* 1, 2
(1996): 131–53.

22 Ian Cook, Philip Crang, and Mark Thorpe, "Tropics of Consumption: 'Getting
with the Fetish' of 'Exotic' Fruit?" in *Geographies of Commodity Chains*, ed. Alex
Hughes and Suzanne Reimer (London: Routledge, 2004), 173–92.

23 Susanne Freidberg, "The Ethical Complex of Corporate Food Power," *Environment and Planning D: Society and Space* 22, 4 (2004): 513–31.

24 For example, in the late 1990s the Council of Canadians and Greenpeace had
been big proponents of mandatory labelling of GMOs (which was under consideration, but would eventually be defeated). Most of the general farm organizations supported only voluntary labelling, arguing that the cost of mandatory
labelling would be downloaded onto them.

25 This statistic increased as the CWB did more consultations with buyers and
as the debate heated up. By early 2003 the CWB was announcing that over 80
percent of Canada's customers for Canada Western Red Spring wheat would
not buy from Canada if it were also growing GM wheat.

26 For example, an Ipsos-Reid poll of market trends and food choices in 2002 found that 58 percent of Canadians thought that using GMOs was a negative trend. This indicated an increase in consumer wariness of GMOs, since only 45 percent identified the use of GMOs in food as negative in 1998. Source: Barbara Duckworth, "People Question Safety of Food," *Western Producer,* 8 October 2004.

27 Council of Canadians testimony to the Senate Standing Committee on Agriculture and Forestry, 8 November 2001. In *Proceedings of the Standing Senate Committee on Agriculture and Forestry,* http://www.parl.gc.ca/Content/SEN/Committee/371/agri/20ev-e.htm?Language=E&Parl=37&Ses=1&comm_id=2

28 Rod MacRae, Holly Penfound, and Charles Margulis, *Against the Grain: The Threat of Genetically Engineered Wheat* (Greenpeace, 2002).

29 The two studies were: C. Andy King, Larry C. Purcell, and Earl D. Vories, "Plant Growth and Nitrogenase Activity of Glyphosate-Tolerant Soybean in Response to Foliar Glyphosate Applications," *Agronomy Journal* 93, 1 (2001): 179–86; Robert J. Kremer, Pat A. Donald, Armon J. Keaster, and Harry Minor, "Herbicide Impact on Fusarium Spp. and Soybean Cyst Nematode in Glyphosate-Tolerant Soybean," Abstract, *Annual Meetings Abstracts of the American Society of Agronomy, Crop Science Society of America and the Soil Science Society of America* (2000), 257.

30 Kinchy, *Seeds, Science.*

31 Canadian Food Inspection Agency, *Detailed Table for 2002 Confined Field Trials,* 2002, http://www.inspection.gc.ca/english/plaveg/bio/dt/dt_02e.shtml (accessed 20 September 2006).

32 Quoted in Barry Wilson, "CFIA Walks Tightrope on GM Wheat Crop Trials," *Western Producer,* 13 March 2003.

33 Fittingly, the details of the contract between AAFC and Monsanto have never been released. It is known that "AAFC provided non-exclusive access to developmental germplasm so that the company could integrate its own proprietary technology." Government of Canada, *Response of the Federal Departments and Agencies to the Petition Filed August 14, 2003 by Greenpeace Canada under the Auditor General Act: Genetically Engineered Wheat: The Precautionary Principle, Biosafety and the Future of Canada's Agriculture,* 2003, http://www.oagbvg.gc.ca/domino/petitions.nsf/viewe1.0/3EEF96F9AFC918F485256E1C0059F43A (accessed 10 September 2006). According to the AAFC scientists I interviewed, they also contributed scientists and lands for field trials, and they stood to gain marginal royalties once the technology was released.

34 See for example Murdoch, Marsden, and Banks, "Quality, Nature."

35 Agricultural Producers Association of Saskatchewan testimony to the House Standing Committee on Agriculture and Agri-Food, 5 June 2003. http://www.parl.gc.ca/HousePublications/Publication.aspx?DocId=968241&Language=E&Mode=1&Parl=37&Ses=2.

36 Saskatchewan Association of Rural Municipalities testimony to the House Standing Committee on Agriculture and Agri-Food, 5 June 2003. http://www.parl.gc.ca/HousePublications/Publication.aspx?DocId=968241&Language=E&Mode=1&Parl=37&Ses=2.

37 Neil Smith, *Uneven Development: Nature, Capitalism, and the Production of Space,* 2nd ed. (Oxford: Basil Blackwell, 1984).

38 William Cronon, "The Trouble with Wilderness: Or, Getting Back to the Wrong Nature," In *Uncommon Ground: Toward Reinventing Nature,* ed. Cronon (New York: Norton, 1995), 69–90.

39 Greenpeace testimony to the Senate Standing Committee on Agriculture and Forestry, 8 November 2001, in *Proceedings of the Standing Senate Committee on Agriculture and Forestry,* http://www.parl.gc.ca/Content/SEN/Committee/371/agri/20ev-e.htm?Language=E&Parl=37&Ses=1&comm_id=2.

40 Macrae, Penfound, and Margulis, *Against the Grain.*

41 Buttel, "Environmental and Post-Environmental."

42 Ibid.

43 Harriet Friedmann, "From Colonialism to Green Capitalism: Social Movements and Emergence of Food Regimes," in *New Directions in the Sociology of International Development,* ed. Frederick H. Buttel and Philip D. McMichael (Amsterdam: Elsevier, 2005), 227–64.

CHAPTER 5: ONLY CONSUMERS GET TO DECIDE IN THE MARKET

1 Mitchell Dean, *Governmentality: Power and Rule in Modern Society* (London: Sage Publications, 1999), 16.

2 Nikolas Rose, *Powers of Freedom: Reframing Political Thought* (Cambridge: Cambridge University Press, 1999), 87.

3 Julie Guthman, "Neoliberalism and the Making of Food Politics in California," *Geoforum* 39, 3 (2008): 1176.

4 Tim Lang and Yiannis Gabriel, "A Brief History of Consumer Activism," in *The Ethical Consumer,* ed. Rob Harrison, Terry Newholm, and Deirdre Shaw (London: Sage, 2005), 39–53.

5   David Blanke, *Sowing the American Dream: How Consumer Culture Took Root in the Rural Midwest* (Athens, OH: Ohio University Press, 2000).

6   Ronald R. Kline, *Consumers in the Country: Technology and Social Change in Rural America* (Baltimore: Johns Hopkins University Press, 2000).

7   Rational and social choice theories are not the same. Dryzek is careful to clarify that social choice theory does not make the behavioural assumption of rational choice theory, namely that individuals act strategically and in a goal-seeking manner. Social choice theory is more concerned with the mechanisms for aggregating individual preferences than the process through which preferences are made.

8   John Dryzek, *Deliberative Democracy and Beyond: Liberals, Critics, Contestations* (Oxford: Oxford University Press, 2000), 34.

9   See for example Trevor J. Barnes, "Rationality and Relativism in Economic Geography: An Interpretive Review of the Homo Economicus Assumption," *Progress in Human Geography* 12, 4 (1988): 473–96; Euclid Tsakalotos, "Homo Economicus, Political Economy and Socialism," *Science and Society* 68, 2 (2004): 137–60.

10 Rose, *Powers of Freedom*, 134.

11 See for example Josée Johnston's analysis of the "citizen-consumer" hybrid as it is embodied by consumers shopping at the food retailer Whole Foods: Josée Johnston, "The Citizen-Consumer Hybrid: Ideological Tensions and the Case of Whole Foods Market," *Theory and Society* 37, 3 (2008): 229–70. Robin Jane Roff examines anti-GM food activism in the U.S. and argues that focussing on individual consumption habits shifts responsibility away from the state and food manufacturers, opens up new markets for business, and does not challenge the increasing prominence of convenience and processed foods: Roff, "Shopping" (see Introduction, n. 4). Julie Guthman makes similar arguments relating to the celebration of organic food consumption: Julie Guthman, "Fast Food/ Organic Food: Reflexive Tastes and the Making of 'Yuppie Chow,'" *Social and Cultural Geography* 4, 1 (2003): 45–58; Julie Guthman, "The 'Organic Commodity' and Other Anomalies in the Politics of Consumption," in *Geographies of Commodity Chains*, ed. Alex Hughes and Suzanne Reimer (London: Routledge, 2004), 233–49.

12 Rachel Slocum, "Consumer Citizens and the Cities for Climate Protection Campaign," *Environment and Planning A* 36, 5 (2004): 763–82; John Clarke and Janet Newman, "What's in a Name? New Labour's Citizen-Consumers and the Remaking of Public Services," *Cultural Studies* 21, 4 (2007): 738–57.

13 Rose, *Powers of Freedom*, 141–42.

14 Ibid., 142; Dean, *Governmentality*, 57.

15 Rajeev Patel, Robert J. Torres, and Peter Rosset, "Genetic Engineering in Agriculture and Corporate Engineering in Public Debate: Risk, Public Relations, and Public Debate over Genetically Modified Crops," *International Journal of Occupational and Environmental Health* 11, 4 (2005): 428–36.

16 The CFIA's decision-making process regarding environmental and health safety is not without controversy. In February 2000 Environment Canada, Health Canada, and the CFIA requested that the Royal Society of Canada (Canada's senior national body of pre-eminent scientists and scholars) convene an expert panel on the future of food biotechnology. This panel evaluated the Canadian regulatory system and its capacity to cope with future products of biotechnology. It found that the regulatory approach that was in place (based on the principle of "substantial equivalence") was not sufficiently precautionary and that the regulatory system was not adequately transparent and open to public scrutiny.

17 Warick, Jason. "Lining Up against GM Wheat: Farmers, Canadian Wheat Board Largely United in Opposing Monsanto's Application to Market 'Roundup Ready' Genetically-Modified Wheat," *Saskatoon StarPhoenix*, August 9, 2003.

18 Murray Fulton, Hartley Furtan, Richard Grey, and George Khachatourians, "Genetically Modified Wheat," *Globe and Mail*, August 11, 2003.

19 Barry Wilson, "Consumers Demand GM labels," *Western Producer*, 11 December 2003.

20 Sean Pratt, "GMO Labelling Fails to Catch On," *Western Producer*, 4 May 2006.

21 See for example Dustin R. Mulvaney, "Identifying Vulnerabilities, Exploring Opportunities: Reconfiguring Production, Conservation, and Consumption in California Rice," *Agriculture and Human Values* 25, 2 (2008): 173–76.

22 Jason Warick, "GM Flax Seed Yanked Off Market: European Customers' Fears Fuel Unprecedented Move," *Saskatoon StarPhoenix*, 22 June 2001.

23 Ibid.

24 Karl Marx, *Grundrisse*, trans. Martin Nicolaus (London: Penguin Books, 1973), 421.

Bibliography

Abergel, Elisabeth, and Katherine Barrett. "Putting the Cart before the Horse: A Review of Biotechnology Policy in Canada." *Journal of Canadian Studies* 37, 3 (2002): 135–61.

Ackroyd, Stephen, and John A. Hughes. *Data Collection in Context.* 2nd ed. London: Longman Group UK, 1992.

Agricultural Producers Association of Saskatchewan testimony to the House Standing Committee on Agriculture and Agri-Food, 5 June 2003. http://www.parl. gc.ca/HousePublications/Publication.aspx?DocId=968241&Language=E& Mode=1&Parl=37&Ses=2.

Agriculture and Agri-Food Canada. *Profile of the Canadian Wheat Industry. Bi-Weekly Bulletin.* 20,5 (2004): 1-8.

Andrée, Peter. "The Biopolitics of Genetically Modified Organisms in Canada." *Journal of Canadian Studies* 37, 3 (2002): 162–91.

——. "The Genetic Engineering Revolution in Agriculture and Food: Strategies of the 'Biotech Bloc.'" In *The Business of Global Environmental Governance*, edited by David L. Levy and Peter J. Newell, 135–166 Cambridge, MA: MIT Press, 2005.

Andrée, Peter, and Lucy Sharratt. *Genetically Modified Organisms and Precaution: Is the Canadian Government Implementing the Royal Society of Canada's Recommendations?* Ottawa: The Polaris Institute, 2004.

Avery, Donald. "The Radical Alien and the Winnipeg General Strike of 1919." In *Canadian Working Class History: Selected Readings,* 3rd ed., edited by Laurel

Sefton Macdowell and Ian Radforth, 217–31. Toronto: Canadian Scholars Press, 2006.

Barnes, Trevor J. "Rationality and Relativism in Economic Geography: An Interpretive Review of the Homo Economicus Assumption." *Progress in Human Geography* 12, 4 (1988): 473–96.

Beckie, H.J., K.N. Harker, L.M. Hall, S.I. Warwick, A. Légère, P.H. Sikkema, G.W. Clayton, et al. "A Decade of Herbicide-Resistant Crops in Canada." *Canadian Journal of Plant Science* 86, 4 (2006): 1243–64.

Berwald, Derek, Colin A. Carter, and Guillaume P. Gruère. "Rejecting New Technology: The Case of Genetically Modified Wheat." *American Journal of Agricultural Economics* 88, 2 (2006): 432–47.

Bjorkquist, Sara, and Mark Winfield. *The Regulation of Agricultural Biotechnology in Canada*. Toronto: Canadian Institute for Environmental Law and Policy, 1999. http://cielap.org/pdf/regbiotch.pdf.

Blanke, David. *Sowing the American Dream: How Consumer Culture Took Root in the Rural Midwest*. Athens, OH: Ohio University Press, 2000.

Boyd, William, W. Scott Prudham, and Rachel A. Schurman. "Industrial Dynamics and the Problem of Nature." *Society and Natural Resources* 14, 7 (2001): 555–70.

Britnell, George Edwin. *The Wheat Economy*. Toronto: University of Toronto Press, 1939.

Bullock, David S., and Elisavet I. Nitsi. "Roundup Ready Soybean Technology and Farm Production Costs: Measuring the Incentive to Adopt Genetically Modified Seeds." *American Behavioral Scientist* 44, 8 (2001): 1283–1301.

Busch, Lawrence, and Arunas Juska. "Beyond Political Economy: Actor Networks and the Globalization of Agriculture." *Review of International Political Economy* 4, 4 (1997): 688–708.

Busch, Lawrence, William B. Lacy, and Jeffrey Burkhardt. *Plants, Power, and Profit: Social, Economic, and Ethical Consequences of the New Biotechnologies*. Cambridge, MA: Blackwell, 1991.

Buttel, Frederick H. "The Environmental and Post-Environmental Politics of Genetically Modified Crops and Foods." *Environmental Politics* 14, 3 (2005): 309–23.

Callon, Michael. "Some Elements in a Sociology of Translation: Domestication of the Scallops and Fishermen of St. Brieuc Bay." In *Power, Action, Belief: A New Sociology of Knowledge?* edited by John Law. London: Routledge and Kegan Paul, 1986.

Canadian Biotehnology Action Network. "Updates: Victory for Democracy in Canada on Genetic Engineering!" 2 December 2010. http://www.cban.ca/Take-Action/Action-Closed-Bill-C-474.

Canadian Food Inspection Agency. *The Biology of Brassica Napus L.* Ottawa: Plant Biosafety Office, 1994.

——. *Detailed Table for 2002 Confined Field Trials.* 2002. http://www.inspection.gc.ca/english/plaveg/bio/dt/dt_02e.shtml (accessed 20 September 2006).

——. "Regulating 'Novelty' and Plants with Novel Traits." http://www.inspection.gc.ca/english/plaveg/bio/pub/novnoue.shtml (accessed 22 November 2011).

Canadian Wheat Board. "About Us: History." http://www.cwb.ca/public/en/about/history/ (accessed 12 December 2007).

Canola Council of Canada. "About Us." http://www.canola-council.org/council.html (accessed 22 January 2008).

Carew, Richard. "Science Policy and Agricultural Biotechnology in Canada." *Review of Agricultural Economics* 27, 3 (2005): 300–16.

Carter, Sarah. "Two Acres and a Cow: 'Peasant' Farming for the Indians of the Northwest, 1889–97." *Canadian Historical Review* 70, 1 (1989): 27–52.

Chayanov, Alexander V. *The Theory of Peasant Economy*, edited by Daniel Thorner, Basil Kerblay, and R.E.F. Smith. Homewood, IL: Richard D. Irwin, 1966.

Clapp, Jennifer. "Transnational Corporate Interests and Global Environmental Governance: Negotiating Rules for Agricultural Biotechnology and Chemicals." *Environmental Politics* 12, 4 (2003): 1–23.

Clarke, John, and Janet Newman. "What's in a Name? New Labour's Citizen-Consumers and the Remaking of Public Services." *Cultural Studies* 21, 4 (2007): 738–57.

Cook, Ian, Philip Crang, and Mark Thorpe. "Tropics of Consumption: 'Getting with the Fetish' of 'Exotic' Fruit?" In *Geographies of Commodity Chains*, edited by Alex Hughes and Suzanne Reimer, 173–92. London: Routledge, 2004.

Cook, Ian, and Philip Crang. "The World on a Plate: Culinary Culture, Displacement and Geographical Knowledges." *Journal of Material Culture* 1, 2 (1996): 131–53.

Cronon, William. "The Trouble with Wilderness: Or, Getting Back to the Wrong Nature." In *Uncommon Ground: Toward Reinventing Nature*, edited by Cronon, 69–90. New York: Norton, 1995.

Council of Canadians. 8 November 2001. "Proceedings of the Standing Senate Committee on Agriculture and Forestry." http://www.parl.gc.ca/Content/ SEN/Committee/371/agri/20ev-e.htm?Language=E&Parl=37&Ses=1&co mm_id=2

Dant, Tim. *Knowledge, Ideology and Discourse: A Sociological Perspective*. London: Routledge, 1991.

Dean, Mitchell. *Governmentality: Power and Rule in Modern Society*. London: Sage Publications, 1999.

Donald, Betsy, and Alison Blay-Palmer. "The Urban Creative-Food Economy: Producing Food for the Urban Elite or Social Inclusion Opportunity?" *Environment and Planning A* 38, 10 (2006): 1901–20.

Downey, R., and H.J. Beckie. *Isolation Effectiveness in Canola Pedigree Seed Production*. Saskatoon, SK: Agriculture and Agri-Food Canada, 2002.

Dryzek, John. *Deliberative Democracy and Beyond: Liberals, Critics, Contestations*. Oxford: Oxford University Press, 2000.

Duckworth, Barbara. "People Question Safety of Food." *Western Producer*, 8 October 2004.

Eisler, Dale. *False Expectations: Politics and the Pursuit of the Saskatchewan Myth*. Regina: Canadian Plains Research Centre, 2006.

Enoch, Simon. *The Potemkin Corporation: Corporate Social Responsibility, Public Relations and Crises of Democracy and Ecology*. PhD diss., Ryerson University and York University, 2009.

Fairbairn, Brett. "Saskatchewan Co-operative Elevator Company. In *Encyclopedia of Saskatchewan*, 802. Regina: Canadian Plains Research Centre, University of Regina, 2005.

Fairbairn, Garry, *From Prairie Roots: The Remarkable Story of Saskatchewan Wheat Pool*. Saskatoon: Western Producer Prairie Book, 1984.

FitzSimmons, Margaret, and David Goodman. "Incorporating Nature: Environmental Narratives and the Reproduction of Food." In *Remaking Reality: Nature at the Millennium*, edited by Bruce Braun and Noel Castree, 194–220. London: Routledge, 1998.

Fontana, Andrea, and James H. Frey. "The Interview: From Structured Questions to Negotiated Text." In *Handbook of Qualitative Research*, 2nd ed., edited by Norman K. Denzin, and Yvonna. S. Lincoln, 645–72. Thousand Oaks, CA: Sage, 2000.

Fowke, Vernon Clifford. *The National Policy and the Wheat Economy*. (Toronto: University of Toronto Press, 1957).

Freidberg, Susanne. "The Ethical Complex of Corporate Food Power." *Environment and Planning D: Society and Space* 22, 4 (2004): 513–31.

Friedmann, Harriet. "From Colonialism to Green Capitalism: Social Movements and Emergence of Food Regimes." In *New Directions in the Sociology of International Development,* edited by Frederick H. Buttel and Philip D. McMichael, 227–64. Amsterdam: Elsevier, 2005.

——. "Household Production and the National Economy: Concepts for the Analysis of Agrarian Formations." *Journal of Peasant Studies* 7, 2 (1980): 158–84.

——. "World Market, State, and Family Farm: Social Bases of Household Production in the Era of Wage Labor." *Comparative Studies in Society and History* 20, 4 (1978): 545–86.

Friesen, Lyle F., Alison G. Nelson, and Rene C. Van Acker. "Evidence of Contamination of Pedigreed Canola *(Brassica Napus)* Seedlots in Western Canada with Genetically Engineered Herbicide Resistance Traits." *Agronomy Journal* 95, 6 (2003): 1342–47.

Fulton, M., H. Furtan, D. Gosnell, R. Gray, K. Giannakas, J. Hobbs, J Holzman, et al., eds. *Transforming Agriculture: The Benefits and Costs of Genetically Modified Crops.* Prepared for the Canadian Biotechnology Advisory Committee Project Steering Committee on the Regulation of Genetically Modified Foods No. 138. Saskatoon: University of Saskatchewan, 2001.

Fulton, Murray, Hartley Furtan, Richard Grey, and George Khachatourians. "Genetically Modified Wheat." *Globe and Mail*, 11 August 2003.

Fulton, Murray, and Lynette Keyowski. "The Producer Benefits of Herbicide-Resistant Canola." *AgBioForum* 2, 1 (1999): 85–93.

Goldsmith, Peter D. "Innovation, Supply Chain Control, and the Welfare of Farmers: The Economics of Genetically Modified Seeds." *American Behavioral Scientist* 44, 8 (2001): 1302–26.

Goodman, David. "Ontology Matters: The Relational Materiality of Nature and Agro-Food Studies." *Sociologia Ruralis* 41, 2 (2001): 182–200.

Goodman, David, Bernardo Sorj, and John Wilkinson. *From Farming to Biotechnology: A Theory of Agro-Industrial Development.* Oxford: Basil Blackwell, 1987.

Goodman, David, and Melanie DuPuis. "Knowing Food and Growing Food: Beyond the Production-Consumption Debate in the Sociology of Agriculture." *Sociologia Ruralis* 42, 1 (2002): 5–22.

Government of Canada. *Response of the Federal Departments and Agencies to the Petition Filed August 14, 2003 by Greenpeace Canada under the Auditor General Act: Genetically Engineered Wheat: The Precautionary Principle, Biosafety and the Future of Canada's Agriculture.* 2003. http://www.oagbvg.gc.ca/domino/petitions.nsf/viewe1.0/3EEF96F9AFC918F485256E1C0059F43A (accessed 10 September 2006).

Greenpeace testimony to the Senate Standing Committee on Agriculture and Forestry, 8 November 2001. In Proceedings of the Standing Senate Committee on Agriculture and Forestry. http://www.parl.gc.ca/Content/SEN/Committee/371/agri/20ev-e.htm?Language=E&Parl=37&Ses=1&comm_id=2

Guthman, Julie. "Fast Food/Organic Food: Reflexive Tastes and the Making of 'Yuppie Chow.'" *Social and Cultural Geography* 4, 1 (2003): 45–58.

——. "Neoliberalism and the Making of Food Politics in California." *Geoforum* 39, 3 (2008): 1171–83.

——. "The 'Organic Commodity' and Other Anomalies in the Politics of Consumption." In *Geographies of Commodity Chains*, edited by Alex Hughes and Suzanne Reimer, 233–49. London: Routledge, 2004.

Hall, Clare. "Identifying Farmer Attitudes Towards Genetically Modified (GM) Crops in Scotland: Are They Pro- or Anti-GM?" *Geoforum* 39, 1 (2008): 204 12.

Hall, Clare, and Dominic Moran, "Investigating GM Risk Perceptions: A Survey of Anti-GM and Environmental Campaign Group Members." *Journal of Rural Studies* 22, 1 (2006): 29–37.

Haraway, Donna. *The Haraway Reader.* New York: Routledge, 2004.

——. "Nature, Politics, and Possibilities: A Debate and Discussion with David Harvey and Donna Haraway." *Environment and Planning D: Society and Space* 13, 5 (1995): 507–27.

——. *Simians, Cyborgs, and Women: The Reinvention of Nature.* New York: Routledge, 1991.

——. *When Species Meet.* Minneapolis: University of Minnesota Press, 2008.

Hartley, Sarah, and Grace Skogstad. "Regulating Genetically Modified Crops and Foods in Canada and the United Kingdom: Democratizing Risk Regulation." *Canadian Public Administration* 48,3 (2005): 305–27.

Hartsock, Nancy. "Rethinking Modernism: Minority versus Majority Theories." *Cultural Critique* 7 (1987): 187–206.

Harvey, David. *A Brief History of Neoliberalism.* New York: Oxford University Press, 2005.

Heller, Chaia. "Post-Industrial 'Quality Agricultural Discourse': Techniques of Governance and Resistance in the French Debate over GM crops." *Social Anthropology* 14, 3 (2006): 319–34.

Henderson, George L. *California and the Fictions of Capital*. New York: Oxford University Press, 1999.

Howlett, Michael, and Andrea Migone. "Explaining Local Variation in Agri-Food Biotechnology Policies: 'Green' Genomics Regulation in Comparative Perspective." *Science and Public Policy* 37, 10 (2010): 781–95.

Ilbery, Brian, and Moya Kneafsey. "Product and Place: Promoting Quality Products and Services in the Lagging Regions of the European Union." *European Urban and Regional Studies* 5, 4 (1998): 329–41.

Jessop, Bob. "Liberalism, Neoliberalism, and Urban Governance: A State-Theoretical Perspective." *Antipode* 34, 3 (2002): 452–71.

Johnston, Josée. "The Citizen-Consumer Hybrid: Ideological Tensions and the Case of Whole Foods Market." *Theory and Society* 37, 3 (2008): 229–70.

Kautsky, Karl. *The Agrarian Question*. Translated by Peter Burgess. London: Zwan Publications, 1988.

Khachatourians, George G., Arthur K. Sumner, and Peter W.B. Phillips. "An Introduction to the History of Canola and the Scientific Basis for Innovation." In Phillips and Khachatourians, *Biotechnology Revolution*, 33–48.

Kinchy, Abby. *Seeds, Science and Struggle: The Global Politics of Transgenic Crops*. Cambridge, MA: MIT Press, 2012.

King, C. Andy, Larry C. Purcell, and Earl D. Vories. "Plant Growth and Nitrogenase Activity of Glyphosate-Tolerant Soybean in Response to Foliar Glyphosate Applications." *Agronomy Journal* 93, 1 (2001): 179–86.

Kleinman, Daniel Lee, and Abby J. Kinchy. "Against the Neoliberal Steam Roller? The Biosafety Protocol and the Social Regulation of Agricultural Biotechnologies." *Agriculture and Human Values* 24, 2 (2007): 195–206.

Kline, Ronald R. *Consumers in the Country: Technology and Social Change in Rural America*. Baltimore: Johns Hopkins University Press, 2000.

Kloppenburg, Jack Ralph. *First the Seed: The Political Economy of Plant Biotechnology*. 2nd ed. Madison: University of Wisconsin Press, 2004.

Kneen, Brewster. *The Rape of Canola*. Toronto: NC Press, 1992.

Kremer, Robert J., Pat A. Donald, Armon J. Keaster, and Harry Minor. "Herbicide Impact on Fusarium Spp. and Soybean Cyst Nematode in Glyphosate-Tolerant Soybean." Abstract. *Annual Meetings Abstracts of the American Society of Agronomy, Crop Science Society of America and the Soil Science Society of America,* 257 (2000). http://www.biotech-info.net/fungi_buildup_abstract.html.

Lang, Tim, and Yiannis Gabriel. "A Brief History of Consumer Activism." In *The Ethical Consumer,* edited by Rob Harrison, Terry Newholm, and Deirdre Shaw, 39–53. London: Sage, 2005.

Latour, Bruno. *We Have Never Been Modern.* Cambridge, MA: Harvard University Press, 1993.

——. "When Things Strike Back: A Possible Contribution of 'Science Studies' to the Social Sciences." *British Journal of Sociology* 51, 1 (2000): 107–23.

Law, John, ed. *A Sociology of Monsters: Essays on Power, Technology and Domination.* London: Routledge, 1991.

Lenin, Vladimir Ilyich. *The Development of Capitalism in Russia.* Moscow: Progress Publishers, 1967.

Lipset, S. M. *Agrarian Socialism: The Cooperative Commonwealth Federation in Saskatchewan, A Study in Political Sociology.* Berkeley: University of California Press, 1950.

MacRae, Rod, Holly Penfound, and Charles Margulis. *Against the Grain: The Threat of Genetically Engineered Wheat.* Greenpeace, 2002.

Magnan, André. "Refeudalizing the Public Sphere: 'Manipulated Publicity' in the Canadian Debate on GM Foods." *Canadian Journal of Sociology* 31, 1 (2006): 25–53.

——. "Strange Bedfellows: Contentious Coalitions and the Politics of GM Wheat." *The Canadian Review of Sociology and Anthropology* 44, 2 (2007): 289–317.

Mann, Susan. *Agrarian Capitalism in Theory and Practice.* Chapel Hill: University of North Carolina Press, 1990.

Mann, Susan A., and James M. Dickinson. "Obstacles to the Development of a Capitalist Agriculture." *Journal of Peasant Studies* 5,4 (1978): 466–81.

Marsden, Terry, and Everard Smith. "Ecological Entrepreneurship: Sustainable Development in Local Communities through Quality Food Production and Local Branding." *Geoforum* 36, 4 (2005): 440–51.

Marx, Karl. *Grundrisse.* Translated by Martin Nicolaus. London: Penguin Books, 1973.

May, Tim. *Social Research: Issues, Methods and Process.* Buckingham: Open University Press, 1993.

Mbengue, Makane Moïse, and Urs P. Thomas. "The Precautionary Principle: Torn between Biodiversity, Environment-Related Food Safety and the WTO." *International Journal of Global Environmental Issues* 5, 1–2 (2005): 36–53.

Mitchell, Timothy. *Rule of Experts: Egypt, Techno-Politics, Modernity.* Berkeley: University of California Press, 2002.

Monsanto Company. "Monsanto to Realign Research Portfolio: Development of Roundup Ready Wheat Deferred." News release. 10 May 2004. http://monsanto.mediaroom.com/index.php?s=43&item=241.

Monsanto Company, "Putting Technology to Work in the Field" 2005 http://www.monsanto.co.uk/news/product_pipeline/pipeline_brochure.pdf

Moore, Elizabeth. "The New Direction of Federal Agricultural Research in Canada: From Public Good to Private Gain?" *Journal of Canadian Studies* 37, 3 (2002): 112–34.

Müller, Birgit. "Infringing and Trespassing Plants: Patented Seeds at Dispute in Canada's Courts." *Focaal European Journal of Anthropology* 48, 1 (2006): 83–98.

Mulvaney, Dustin R. "Identifying Vulnerabilities, Exploring Opportunities: Reconfiguring Production, Conservation, and Consumption in California Rice." *Agriculture and Human Values* 25, 2 (2008): 173–76.

Murdoch, Jonathan. "Inhuman/Nonhuman/Human: Actor-Network Theory and the Prospects for a Nondualistic and Symmetrical Perspective on Nature and Society." *Society and Space* 15, 6 (1997): 731–56.

Murdoch, Jonathan, Terry Marsden, and Jo Banks. "Quality, Nature, and Embeddedness: Some Theoretical Considerations in the Context of the Food Sector." *Economic Geography* 76, 2 (2000): 107–25.

Newell, Peter, and Ruth Mackenzie. "Whose Rules Rule? Development and the Global Governance of Biotechnology." *IDS Bulletin* 35, 1 (2004): 82–91.

Patel, Rajeev, Robert J. Torres, and Peter Rosset. "Genetic Engineering in Agriculture and Corporate Engineering in Public Debate: Risk, Public Relations, and Public Debate over Genetically Modified Crops." *International Journal of Occupational and Environmental Health* 11, 4 (2005): 428–36.

Peck, Jamie, and Adam Tickell. "Neoliberalizing Space." *Antipode* 34, 3 (2002): 380–404.

Phillips, Peter W.B. "The Role of Private Firms." In Phillips and Khachatourians, *Biotechnology Revolution*, 105–28.

Phillips, Peter W.B., and George G. Khachatourians, eds. *The Biotechnology Revolution in Global Agriculture: Invention, Innovation and Investment in the Canola Sector.* New York: CABI Publishing, 2001.

———. "Introduction and Overview." In Phillips and Khachatourians, *Biotechnology Revolution*, 3–20.

———. "The Role of Public Sector Institutions." In Phillips and Khachatourians, *Biotechnology Revolution*, 105–28.

Pratt, Sean. "GMO Labelling Fails to Catch On." *Western Producer,* May 4, 2006.

Prudham, Scott, and Angela Morris. "Regulating the Public to Make the Market 'Safe' for GM Foods: The Case of the Canadian Biotechnology Advisory Committee." *Studies in Political Economy* 78 (2006): 145–75.

Purdue, Derrick. A. *Anti-GenetiX: The Emergence of the Anti-GM Movement.* Aldershot: Ashgate, 2000.

Quijano, Romeo. F. "Elements of the Precautionary Principle." In *Precaution, Environmental Science, and Preventive Public Policy*, edited by Joel A. Tickner, 21–28. Washington: Island Press, 2003.

Reisner, Ann Elizabeth. "Social Movement Organizations' Reactions to Genetic Engineering in Agriculture." *American Behavioural Scientist* 44, 8 (2001): 1389–1404.

Renting, Henk, Terry K. Marsden, and Jo Banks. "Understanding Alternative Food Networks: Exploring the Role of Short Food Supply Chains in Rural Development." *Environment and Planning A* 35 (2003): 393–411.

Roff, Robin Jane. "Shopping for Change? Neoliberalizing Activism and the Limits to Eating Non-GMO." *Agriculture and Human Values* 24, 4 (2007): 511–22.

Rose, Nikolas. *Powers of Freedom: Reframing Political Thought.* Cambridge: Cambridge University Press, 1999.

Saskatchewan Association of Rural Municipalities testimony to the House Standing Committee on Agriculture and Agri-Food, June 5, 2003. http://www.parl. gc.ca/HousePublications/Publication.aspx?DocId=968241&Language=E& Mode=1&Parl=37&Ses=2.

Schurman, Rachel. "Fighting Frankenfoods: Industry Structures and the Efficacy of the Anti-Biotech Movement in Western Europe." *Social Problems* 51, 2 (2004): 243–68.

Schurman, Rachel, and William Munro. "Ideas, Thinkers, and Social Networks: The Process of Grievance Construction in the Anti-Genetic Engineering Movement." *Theory and Society* 35, 1 (2006): 1–38.

Sharp, Paul F. *The Agrarian Revolt in Western Canada.* 1948. Reprint, Regina: Canadian Plains Research Centre, University of Regina, 1997.

Shiva, Vandana. *Biopiracy: The Plunder of Nature and Knowledge.* Cambridge, MA: South End Press, 1997.

Slocum, Rachel. "Consumer Citizens and the Cities for Climate Protection Campaign." *Environment and Planning A* 36, 5 (2004): 763–82.

Smith, Neil. *Uneven Development: Nature, Capitalism, and the Production of Space.* 2nd ed. Oxford: Basil Blackwell, 1984.

Statistics Canada. "Farm Data and Farm Operator Data: 2006 Census of Agriculture." 1 December 2008. http://www.statcan.gc.ca/pub/95-629-x/95-629-x2007000-eng.htm.

Symko, Stephan. *From a Single Seed: Tracing the Marquis Wheat Success Story in Canada to Its Roots in Ukraine.* Ottawa, Ontario: Research Branch Agriculture and Agri-Food Canada, 1999.

Thomas, Julian. "A Technical Critique of the Western Canada Quality Assurance (QA) System in Wheat." Unpublished manuscript, 2006.

Tickner, Joel A. "The Role of Environmental Science in Precautionary Decision Making." In *Precaution, Environmental Science, and Preventive Public Policy*, edited by Tickner, 3–20. Washington: Island Press, 2003.

Tokar, Brian. "A history of biotech giant Monsanto." CBC digital archives, 3 May 1999. http://www.cbc.ca/archives/categories/economy-business/agriculture/genetically-modified-food-a-growing-debate/history-of-biotech-giant-monsanto.html.

Tsakalotos, Euclid. "Homo Economicus, Political Economy and Socialism." *Science and Society* 68, 2 (2004): 137–60.

Turner, R. Steven. "Of Milk and Mandarins: RBST, Mandated Science and the Canadian Regulatory Style." *Journal of Canadian Studies* 36, 3 (2001): 107–30.

Varty, John F. "On Protein, Prairie Wheat, and Good Bread: Rationalizing Technologies and the Canadian State, 1912–1935." *Canadian Historical Review* 85, 4 (2004): 721–53.

Waiser, Bill. *Saskatchewan: A New History* (Calgary: Fifth House, 2005).

Warick, Jason. "GM Flax Seed Yanked Off Market: European Customers' Fears Fuel Unprecedented Move." *Saskatoon StarPhoenix,* June 22, 2001.

———. "Lining Up against GM Wheat: Farmers, Canadian Wheat Board Largely United in Opposing Monsanto's Application to Market 'Roundup Ready' Genetically-Modified Wheat." *Saskatoon StarPhoenix*, 9 August 2003.

Warnock, John W. *Saskatchewan: The Roots of Discontent and Protest.* Montreal: Black Rose Books, 2004.

Weiss, Robert S. *Learning from Strangers: The Art and Method of Qualitative Interview Studies.* New York: The Free Press, 1994.

Whatmore, Sarah. *Hybrid Geographies: Natures, Cultures, Spaces*. London: Sage Publications, 2002.

Wilson, Barry. "CFIA Walks Tightrope on GM Wheat Crop Trials." *Western Producer,* 13 March 2003.

——. "Consumers Demand GM labels." *Western Producer*, 11 December 2003.

——. "Gov't Limits Debate on CWB Bill." *Western Producer Daily News*, 20 October 2011.

——. "Labeling Bill Misses Automatic Vote Rule." *Western Producer*, 28 November 2002.

——. "Market Risk Once Part of Process." *Western Producer,* 10 April 2003.

——. "Vanclief Floats Nonscientific GM Test." *Western Producer,* 30 October 2003.

Index

B

barley, 66, 99, 154n26, 157n31

Barrett, Katherine, 31, 35

Bayer Crop Science, 9, 112, 144; SOD's court case against, 17, 102–3

Bill C-18, *Marketing Freedom for Grain Farmers Act,* 46

Bill C-474, *An Act Respecting the Seeds Regulation,* 47–48

biofuels, 66

biopiracy, 5, 150n5

biotech industry, 146; AAFC's role in, 104; development in Canada, 30–33, 49; investment in Australia, 142

biotech policy and regulation (Canada): advisory committees and panels, 38–42, 50, 165n16; government funding and investment, 30–32; lack of transparency and democracy in, 91, 103–5, 110–11, 114; of PNTs, 35–36, 37; pro-biotech campaigns, 32, 114; promotional approach, 31–34, 42, 49–50; public consultations, 37–38, 40–42; regulatory framework, 33–36, 49–50; science- and product-based approach, 38–39, 42–43, 49–50. *See also* regulation

breeding programs. *See* plant breeding

Buttel, Frederick, 92, 114

C

Caccia, Charles, 47–48, 134

Canada Grain Act (1912), 43

Canadian Biotechnology Action Network, 47, 73, 146

Canadian Biotechnology Advisory Committee (CBAC), 127; public consultations, 40–41

Canadian Census of Agriculture, 28

Canadian Environmental Protection Act (CEPA), 34

Canadian Foodgrains Bank, 72

Canadian Food Inspection Agency (CFIA): conflict, 43, 103, 111; critique of Canadian biotech policy, 39, 42; environmental risk assessment, 40, 126, 165n16; farmers' requests for locations of field trials, 103; interview difficulties with, x; regulation of PNTs, 35–36, 37; support for RR wheat, 125; variety registration process, 45, 64, 112, 128

Canadian General Standards Board, 48, 134

Canadian Grain Industry Working Group on Genetically Modified Wheat, 100

Canadian Health Coalition (CHC), 100, 104

Canadian Wheat Board (CWB), 161n25; dismantling of, 17, 63, 147; establishment, 43–44, 61, 151n5; federal lawsuit, 46; opposition to RR wheat, 44, 99–100, 107, 131; plebiscite on marketing of barley, 154n26, 157n31; role of, 10, 62–65, 89; single-desk structure, 44, 46, 63, 154n26 canola: associated with innovation, 73, 82–83, 86–88; biological and agronomic characteristics of, 83–85; breeding of, 77, 80–84; as a companion species, 55, 73, 88; contamination of, 85, 102–3, 132–33; crop rotation, 68, 83–85; cultivars, 65, 85; cultural identity of, 72–73, 87–88; herbicide-tolerant varieties, 3, 51, 69, 85, 102; hybrid varieties, 82, 129, 131; materiality and semiotics of, 54–55, 89; political organizing around, 76–77; price, 76, 78; private investment in, 77, 79–81, 83–86, 89; promiscuous nature of, 83, 85; public awareness of GM varieties, 104; seeded areas in the prairie provinces, 57, 77; seeds, 78, 82–85, 101–3, 131; social relations of, 52; success of GM varieties, 19, 87, 117; transformation of, 52. *See also* rapeseed

Canola Council of Canada (CCC), 86–97, 130, 139; mission, 77; research funding program, 80; support for RR wheat, 18–19, 125

capital: global, 6–7; industrial, 26–27; and labour, 140; nature as an obstacle to, 25–27, 29

capitalism: and agriculture, 22, 25–28; structural relations of power and, 54; and transformation of the countryside, 23–24; and value of commodities, 29

Chayanov, Alexander, 24–25, 29

citizenship, 123–24

classes. *See* social classes

coalition against RR wheat: access to information requests, 104–5; arguments for opposition, 18, 90–91, 99–105, 112–13, 128–35; collective action, 135–39, 141, 147; and consumer choice, 117–18; interview methods, 13, 16; organizations involved, 5–6, 8, 14, 16–17; struggles and differences within, 17–18, 97–98; success, 142. *See also individual organization*

collectivity and collective action, 118, 123, 135–39, 141, 147

commercialization: of GM flax, 139; of GM products in Canada, 30, 33, 40, 42; role of private companies in, 31, 80; of RR wheat, 44, 143; synchronized (of wheat), 142–44

commodity chains, 115; wheat, 62, 64, 108, 113

companion species, 54–55, 73, 88

Conservative Party of Canada, 46, 151n5, 157n31

consumers: activism/movements, 5, 93–96, 120–22, 141; choice in the market, 117–24, 139–40; concerns or fears of GM foods, 92, 100, 162n26; "consumer knows best" discourse, 107–8; cooperatives, 121; farmers as, 140–41; mandatory labelling of GM products, 48–49, 51, 104, 134–35; urban, 73

consumption: alternative, 122; discourses and politics of, 93–96, 106, 120; ethics and, 123–24; habits, 164n11. *See also* consumers

contamination: AAFC study on, 104–5, 158n60; canola, 85, 102–3, 132–33, 158n60; flax, 146; GM wheat, 103, 108, 146

cooperatives: consumer, 121, 141; grain elevators, 17, 59–60; insurance, 61; marketing, 52, 59–60; movements in prairie farm history, 136; wheat pools, 17, 60–61, 63, 89

corn, 80, 82, 85

corporate control: of agriculture, 7, 93, 115–16, 145; of family farms, 28, 146–47; of GMOs, 125, 147–48; of seeds, 91, 111, 125

Council of Canadians (CoC), 97–98, 134; support for mandatory labelling, 100, 104, 161n24

Cronon, William, 109

Crop Development Centre, 137–39

Croplife Canada, 125, 129

crop rotation: of canola, 68, 83–85; of wheat, 67–68

D

Dean, Mitchell, 119, 124

democracy: lacking in Canadian biotech strategy, 91, 103–5, 110–11, 114; in the marketplace, 123

Development of Capitalism in Russia (Lenin), 23–24

Dickinson, James, 25–27, 29

dog-human relationships, 54–55

Downey, Keith, 75

Dryzek, John, 123, 164n7

E

economic impact: of GMOs, 43, 104; of RR wheat, 44, 49, 128

environmental and health risks: activism, 90, 92–95, 98; legislation and regulation of, 34, 165n16

environmental safety assessment, 44; performed by GMO manufacturers, 40; on PNTs, 36, 37

European Union: anti-GM movements, 4, 92, 122; approval of GM food, 113; canola research,

87; demand for organic food, 108; precautionary approach to regulation of GMOs, 31, 39; quality food production, 95; rejection of GM foods, 13, 91, 138; wheat exports, 15

F

family farms, 24–25; decline of, 71; threat of corporate takeover, 28, 146–47

farmers: ability to save seed, 5, 28, 66, 82, 129, 131, 146; agency, 140; collective action, 118, 135–39, 141, 147; as consumers, 140–41; entrepreneurial spirit of, 87; French, 94; leadership in anti-GM movements, 4; loans and credit, 27–28; market choices, 126, 129–31; organic, 68, 98, 102, 132–33; political organizing, 58–60, 62, 150n4; private property, 135–36; response to consumer demand, 107–8; royalty payments, 66; shared governance, 44, 63, 151n5; small, 23, 94, 147; threat of global capital, 7. *See also* coalition against RR wheat; National Farmers Union (NFU); producers

flax. *See* GM flax

Flax Council of Canada (FCC), 137, 146

food safety, 95, 115, 126–27

Fowke, Vernon, 56

Friedmann, Harriet, 24–25, 115, 156n14, 156n19

G

genetically modified organisms (GMOs): approval of, 33, 47, 112–13, 142; contamination of non-GM crops, 4, 17, 102–3, 132–33, 146; corporate control of, 125, 147–48; global movements against, 6; labelling of, 48–49, 51, 104, 134–35; market opposition to, 100–101; promotion of, 110–11, 114; risks of, 30, 34, 38, 92–94, 125. *See also* anti-GM movements; genetic engineering (GE); genetic modification (GM)

genetic engineering (GE): equivalence to traditional plant breeding, 34–35, 38; ethical issues, 38; as a novel process, 34; PNTs, 35–36, 37; political debates on, 46–50; securing public acceptance of, 40–42. *See also* Monsanto

genetic modification (GM): academic awareness/debates on, 69, 91–96; benefits and opportunities of, 114, 125, 130; of canola, 83–89; of flax, 135, 137–39, 145–46; impact on organic farming, 102; moral implications of, 5; risky traits, 69, 92. *See also* anti-GM movements; genetically modified organisms (GMOs); GM wheat; Monsanto

glucosinolates, 74, 76

GM alfalfa, 47

GM flax (Triffid), 135, 137–39, 145–46

GM maize, 113

GM wheat: contamination from, 103, 108, 146; investment and development in, 144–45; market choice in, 117–20, 135; politics of opposition to, 5–6, 145; rejection of, 13, 143–44, 161n25; segregation systems, 100, 145–46;

synchronized commercialization of, 142–44

Goodman, David, 26, 95

governance: neoliberal, xiii, 77, 123–24; shared, 44, 63, 151n5

grain elevators, 43, 58–59, 61; cooperatives, 17, 59–60

Grain Growers' Grain Company (GGGC), 59–60

Grain Growers of Canada, 19, 125–26

grain handling, 58, 67, 147; cooperatives, 59, 61; regulation, 43, 59

Greenpeace: proponent of mandatory labelling, 100–101, 104, 161n24; role in coalition against RR wheat, 6, 90, 97–98, 101–2, 109–10

Grundrisse (Marx), 140

Guthman, Julie, 120, 164n11

H

Haraway, Donna, 52–55

Health Canada, 39, 51, 165n16; novel food assessment, 36, 37

household production, 25, 55, 58, 156n14, 156n19

House of Commons: debates on genetic engineering, 46–48, 50; mandatory labelling bill, 48–49, 51, 134–5; Standing Committee on Agriculture and Agri-Food, 38, 47, 103, 107–8; Standing Committee on Environment and Sustainable Development, 38

human and non-human relations, 19; human/dog co-evolution, 54–55; intentionality and agency, 53–54; representations of hybridity in, 52–53; and social relations, 52, 54

hybrid varieties, 82, 129–31

I

Indigenous peoples, 55, 71

individualism: and market choice, 117–20, 123, 125–28; neoliberalism and, xiii, 119–20; societal structures and, xi–xiv

J

Japan: rapeseed imports, 74, 76; rejection of GM foods, 91; wheat imports, 13, 15

K

Kautsky, Karl, 23–25

Keynesianism, xi, 95, 161n19

Keystone Agricultural Producers (KAP), 16–18, 68, 101, 144, 150n4

Kloppenburg, Jack, 26–27, 80, 82

L

labour: and capital, 140; exploitation, 135; on family farms, 24–29; neoliberalism and, xi

Lenin, Vladimir, 23–25

M

Magnan, André, 42, 94

mandatory labelling, 104, 161n24; private member's bill on, 48–49, 51, 134–35

Manitoba Grain Act (1900), 43, 59

Mann, Susan, 25–27, 29

market acceptance, 99–100, 107, 113

market choice: arguments against, 128–34, 139; and concept of freedom,

140–41; of GM varieties, 130–31; individual, 119, 124–27, 140

market impact, 46–47, 127–28, 131

marketing: of canola and rapeseed, 79, 88; collective, 46, 50, 147; cooperative, 52, 59–60, 137; of wheat, 63, 151n5. *See also* Canadian Wheat Board (CWB)

marketplace, 35, 129; consumer choice in, 118, 123–26; demand, 107, 117

Marx, Karl, 25; *Grundrisse,* 140

Marxism, xii

materiality: of nature, 25, 53; and semiosis, 54; of wheat and canola, 54–55, 89

milk, 51, 72

monopolies, 58, 61. *See also* Canadian Wheat Board (CWB)

Monsanto: appropriationism of, 28; discontinuance of RR wheat, 3, 13, 22, 47, 113; field trials, 90, 103; products and operations, 9–10, 144–45; recombinant bovine growth hormone, 51, 72; Roundup herbicide, 8–9, 45–46, 149n1; SOD's court case against, 17, 102–3. *See also* Roundup Ready wheat (RR wheat)

Moore, Elizabeth, 45, 64–66

N

National Advisory Board on Science and Technology, 32

National Association of Wheat Growers (NAWG), 143

National Biotechnology Advisory Committee (NBAC), 38, 40–41

National Biotechnology Strategy (Canada), 31–33, 40, 42

National Farmers Union (NFU), 104, 108, 111, 131, 139; political history, 17, 98, 136

nationalism: and discourses on canola, 87–88; and narratives of wheat, 70–71

National Policy (1879), 56, 58

National Research Council, 79

nature: as active subject(s), 53; as an obstacle to capital, 25–27, 29; human agency and, 53–54; as separate from humans, 109–10; unevenness of, 23

neoliberalism: and activism, 93–94; and agriculture, xii; definition of, x; governance and, xiii, 77, 123–24; individualism and, xiii, 119–20; policy and practice, x–xi, 45–46; and subjectivity, xiv, 120, 141

North American Millers Association (NAMA), 143

O

oilseed industry, 74, 76, 79

organic farming, 17, 105, 108, 133; inability to grow organic canola, 84, 89, 102, 132; of wheat, 68–69, 105

organizing. *See* political organizing

outcrossing, 69, 85, 103

P

Plant Breeders' Rights (PBR), 66, 79–80

plant breeding: agendas, 63, 66, 76, 131; canola, 77, 80–84; corn, 80, 82; equivalence to GE products, 34–35, 38; experimental farms, 64; rapeseed, 74–76, 79; and reliance on public institutions, 81; in U.S., 27, 82; wheat, 63–68, 157n37

plants with novel traits (PNTs), x, 35–36, 37, 128

political organizing: around canola, 76–77; in Canadian prairie history, 58–60, 62, 150n4; collective action, 135–39, 141, 147

private investment: in canola, 77, 79–81, 83–86, 89; in wheat (lack of), 65–66, 82, 89, 142

private property, x, 27, 135–36

private-sector interests, 33, 45, 50, 105

privatization, xi, 29, 81

producers: activism and collective action, 5, 94, 96, 141, 147; common interests, 136–37; and consumer choice, 107–8; and importance of labour and products, 29; influence in wheat industry, 63, 99; interests and struggles, 43–46, 50, 108, 115; and market choice, 117–18, 127; political organizing, 58–60, 62; research programs, 65–66; vulnerability of, 111, 113. *See also* farmers

promotion of biotechnology: by Canadian state, 31–34, 40, 42, 49–50; of canola, 80; conflict in, 111; of rapeseed, 79

public consultations, 37–38, 104, 114; held by CBAC, 40–41

public investment: in rapeseed, 79; shift to private investment in canola, 77, 80–81; in wheat industry, 62, 65–67

public-private partnerships, xi, 33, 45, 114

public-sector institutions, 50, 65, 81, 85, 105

Q

quality food production, 95–96, 107

Quijano, Romeo, 39

R

railways, 58–59

rapeseed: "double-zero" variety, 76–77, 79; imports and exports, 74, 76; for industrial and dietary purposes, 74; name change, 76; plant breeding, 74–76, 79; production in Canada, 75. *See also* canola

Rapeseed Association of Canada (RAC), 77, 79. *See also* Canola Council of Canada (CCC)

rational and social choice theories, 123, 164n7

recombinant bovine growth hormone, 38, 51, 72

regulation: for commercial interests, 105; of GM flax, 138; of new varieties, 45, 64, 128; participatory approach to, 31; of plants with novel traits (PNTs), x, 35–36, 37, 128; precautionary approach to, 31, 38–39, 41, 165n16; and promotion of GMOs, 111; science- and product-based approach to, 38–39, 42–43, 49–50. *See also* biotech policy and regulation (Canada)

research methods: farm meetings, 22; interview participants, xi–x, 13, 16, 19–22; journalistic sources, 22

Roff, Robin, 93, 164n11

Rose, Nikolas, 117, 119–20, 124

Roundup herbicide, 8–9, 45–46, 149n1

Roundup Ready wheat (RR wheat): AAFC partnership in development of, 104, 162n33; conflicting role of the state over, 111; consumer/individual choice on, 117–20, 125–26, 135; corporate benefits of, 108; cost-benefit analysis, 29, 44; discontinuance of, 3, 13, 22, 47, 113; economic and market impact of, 44, 49, 99, 127–28, 131; environmental impacts of, 101–2, 109; field trials, 90, 103; international rejection of, 91; opposition to, 3–4, 18, 112–14, 118, 128–29; regulatory *vs.* voluntary approach to, 108–9, 126; support for, 18–19, 125–27, 129–30, 140–41, 150n4; threat to existing systems of production, 131–33. *See also* coalition against RR wheat

Royal Society of Canada (RSC), 39–40, 165n16

S

Saskatchewan Agriculture and Food, 16, 86–87, 126

Saskatchewan Association of Rural Municipalities (SARM), 17, 103, 108

Saskatchewan Grain Grower's Association, 59–60

Saskatchewan Organic Directorate (SOD), 68, 72, 88, 104, 111–12, 133; arguments against RR wheat, 102; class action against Monsanto and Bayer, 17, 102–3, 132; differences with conventional farm groups, 18, 98–99

Saskatchewan Wheat Pool (SWP), 61

seeds: canola, 78, 82–85, 101–3, 131; commodification of, 26–27, 145; corn, 82; corporate control of, 91, 111, 125; GM flax, 138–39; hybrid vs. open-pollinated, 129; market choice for GM varieties, 131; private breeding companies, 81; rapeseed, 76; reproduction of, 27–28, 66, 138, 144; rights, 93; sterile, 82

Seeds Act, 34–35, 50, 136

seed saving: of flax, 146; loss of farmer's ability for, 5, 28, 66, 82, 129, 131, 146; Monsanto's restrictions on, 9, 28, 46, 145; practices in wheat, 62, 66, 82, 106; vs. buying of canola, 82–83, 131

settlement of the prairies, 55–56, 58, 62; and cultural importance of wheat, 71; homesteads, 56, 155n13

social classes, 23–24, 124; working class, xiv, 121

social relations, 52, 54, 109, 123

Solutia, 9

"sound science" paradigm, 39, 128

subjectification and subjectivity: concepts, xi, xiii; consumer choice and, 140; liberal, 125; neoliberal, xiv, 120, 141; peasant, 23; political, 139; in researcher-interviewee relationship, 20

substantial equivalence, 35, 37–38, 40, 165n16

symbolism: social relations and, 54; of wheat, 62, 70–73, 88–89

T

Territorial Grain Growers' Association (Saskatchewan Grain Grower's Association), 59–60

"third food regime," 115

U

United States: plant breeding in, 27, 82; wheat imports, 15; wheat production, 11, 12, 71

U.S. Wheat Associates (USWA), 143

V

"value for money" movement, 121–22

Van Acker, Rene, 133

Vanclief, Lyle, 46–47, 103

variety recommending committees, 44–45, 112, 128

voluntary labelling, 48–49, 134–35, 161n24

W

weeds, 69, 84, 101, 110

Western Barley Growers Association, 19, 65, 125, 132

Western Canadian Wheat Growers Association, 19, 142

Western Grain Research Foundation, 66

Western Producer, The, 22, 46, 48

wheat: biological and agronomic characteristics of, 67–70, 157n38; breeding and research, 63–68, 157n37; as a companion species, 55, 88; cultivars, 65; lack of innovation in, 142; materiality and semiotics of, 54–55; quality and protein content, 10, 68, 70–71, 157n34; seed saving practices, 62, 66, 82, 106; social relations of, 52; symbolic and cultural importance of, 62, 70–73, 88–89. *See also* GM wheat; Roundup Ready wheat (RR wheat)

wheat economy: commodity chains, 62, 64, 108, 113; establishment and growth of settler, 55–56, 62, 71; markets, 10, 65, 99, 108, 131, 156n14; national narratives of, 71–72; pools, 17, 60–61, 63, 89; price, 61, 66, 78; producer influence in, 63, 99; public vs. private investment in, 62, 65–67, 77, 82, 89, 142; royalty rates, 66; tariff protection, 56, 58. *See also* Canadian Wheat Board (CWB)

wheat production: distribution areas in Canada, 10, 11; experimental farms, 64; family farms, 24–25; major producers, 11, 12; organic farming, 68–69, 105; in prairie provinces, 56, 57, 156n14; yields, 68. *See also* farmers; producers; wheat economy

When Species Meet (Haraway), 54–55

Wild Rose Agricultural Producers (WRAP), 16, 150n4

Z

zero till practice, 84, 87, 101, 110